ONE
OF
US

ONE OF US

conjoined
twins
and the
future
of
normal

ALICE DOMURAT DREGER

Harvard University Press

Cambridge, Massachusetts

London, England

2004

Pages 190–191 constitute an extension of the copyright page.

Library of Congress Cataloging-in-Publication Data

Dreger, Alice Domurat.
    One of us : conjoined twins and the future of normal / Alice Domurat Dreger.
      p.   cm.
    Includes index.
    ISBN 0-674-01294-1 (alk. paper)
    1. Conjoined twins.   2. Abnormalities, Human.   I. Title.

  RG626.D74 2004
  362.196'043—dc22   2003056974

To Kepler

more attached and more separate each day

Contents

Introduction *1*

1 **The Limits of Individuality** *17*

2 **Split Decisions** *51*

3 **What Sacrifice** *83*

4 **Freeing the Irish Giant** *113*

5 **The Future of Anatomy** *142*

Notes *157*

Acknowledgments *188*

Credits *190*

Index *193*

Illustrations

1. Chang and Eng Bunker as young men   *20*

2. The Bunker twins with two of their sons   *23*

3. Types of conjoinment   *28*

4. Laloo and his parasitic twin   *30*

5. Abigail and Brittany Hensel at play in the family home   *38*

6. Eng and Chang Bunker engaged in various pursuits   *39*

7. Lin and Win Htut before separation   *69*

8. Cover of *AORN Journal,* January 1982   *79*

9. The Two-Headed Boy of Bengal   *114*

10. Charles Byrne with two other giants and several people with dwarfism   *119*

11. Advertising pamphlet for Millie and Christina McCoy   *122*

12. *Crouching Figure with Visible Skeleton,* by Laura Ferguson   *131*

**ONE
OF
US**

Introduction

A pair of conjoined twins walk into a bar. One of them orders a drink from the bartender, who looks the twin over and asks for proof of her age. Hearing the bartender's request, the other twin turns around so that she's the one facing the bartender. Because the second twin appears older, the bartender reconsiders and decides to serve the drink without seeing proof of age.

Another true story: An unrelated man and woman, Americans visiting Japan, try to enter a bar together. But they can't find one that will allow them in. Both are gay and they're in Ni-chome, the "queer" district of Tokyo. The bouncers at the lesbian bars won't let him in, because he's obviously a man. The bouncers at the gay bars won't let her in, because she's obviously a woman.

Another: Three teenagers with dwarfism go into a bar and order drinks. The bartender, too embarrassed to ask for their ID's, goes ahead and serves them. He's afraid to risk offending them by asking them to prove they're legal adults.

And one more: A young woman walks into a bar with a group of friends. Within a few minutes, she thinks she may be going deaf. Everyone around her is communicating as usual, but she can't hear what they're saying above the background noise. Soon she figures it out: it's not that she's going deaf—it's that she's legally blind. She can't read lips and gestures the way others are doing, to "hear" above the noise.

The first story was told to me by Lori Schappell, who is conjoined at one side of her forehead to her sister Reba.[1] The second story was told to me by Cheryl Chase, who was born with a condition known as true hermaphroditism, characterized by mixed sex anatomy (both ovarian and testicular tissue)—not that bar bouncers can tell this sort of thing.[2] The third story was told to me by Danny Black, the owner of ShortDwarf.com, a talent agency and distributor of specialized products for people with small stature.[3] The fourth story came from a medical student named Ruta Sharangpani, who told it to me after I'd said to her, "Hold on a minute. Let me put on my glasses so I can hear you better."[4]

Anatomy matters a lot, and not just in bars. It matters because the senses we possess, the muscles we can control, the resources we require to keep our bodies alive limit and affect what we can experience in any given context: a bar, a school, a house, a courtroom, a subway, a mountaintop, the deep sea, or outer space. But anatomy also matters because it influences the assumptions people make on the basis of our anatomies: that we are too young-looking to drink, too male-looking to be in a lesbian bar, conjoined and therefore incapable of a meaningful, individual life.

Anatomical restrictions have long been explicitly written into the rules governing human life. When kings and queens reigned, their power derived from society's robust notion of how much anatomy mattered: certain people had the birthright. Thousands of years ago, the book of Leviticus stipulated that only men with perfect bodies were worthy of becoming priests: "And the Lord said to Moses, 'Say to Aaron, None of your descendents throughout their generations who has a blemish may approach to offer the bread of his God. For no one who has a blemish shall draw near, a man blind or lame, or one who has a mutilated face or a limb too long, or a man who has an injured foot or an injured hand, or a hunchback, or a dwarf, or a man with a defect in his sight or an itching disease or scabs or crushed testicles'" (21: 16–24). It went without saying that no woman should draw near. The liberal progression started by the Enlightenment has loosened many of these sorts of anatomical rules, at least in the United States. The radical theory that "All men are created equal"—in itself an anatomical claim, though a relatively generous one—eventually gave rise to the practice of allowing people to vote regardless of whether they are men or women, black or white or brown. The legislation known as Title IX (1972 amendments to the Civil Rights Act of 1964) enforced the idea that girls should be as involved in sports as boys. In Vermont, a man can now enter into a civil union with another man.

Because anatomically based rules help to maintain order and protect those perceived as vulnerable, and because restriction brings privilege and privilege is pleasurable, we still have many regulations that dictate who can do what based on anatomy: who can drink alcohol, who can marry whom, who can vote, who can play golf as a member at the Augusta National Club, who may be afforded special legal protections or be promised equity. Beyond the written anatomical rules are the unwritten ones that do the same sort of work of maintaining order, protecting the vulnerable, and restricting privilege. These are the rules—or norms, or standards, call them what you will depending on how stringent they seem at any given moment—that tell us what to expect of a dark-skinned old woman who wears glasses and walks upright, what to expect of a tall light-skinned man with a shaved head and lots of tattoos. We learn and relearn these from our parents, from our peers, from our own bodily experiences, from advertisements, from almost every human encounter.[5] We learn these rules well enough to manipulate our bodies, sometimes slightly and sometimes drastically, to shape the assumptions made about them.

The truth is, most of us go through minor anatomical "normalization" procedures every day, changing our bodies ever so slightly to fit the identity we wish to present socially. We brush the plaque off our teeth, in part to keep them healthy but also so that they won't disgust others with a smell or appearance that would signal we are unclean (and therefore, by the rules of anatomy and identity, slovenly or poor or ill). We wash and style our hair and put on clothes meant to signal who we are underneath (man, woman, corporate team player, professor, artist, rebel). We add a wristwatch to enhance our imperfect internal clocks, to keep our bodily movements well timed in relation to others'. We shave various parts of our bodies depending on what kind of sexuality we wish to signal. We put on eyeglasses or slide into a wheelchair to compensate for the anatomical deficits that might otherwise keep us out of the stream of human life, which largely requires sight and autonomous movement. We worry about getting too fat, knowing that fat is widely equated with weak will and ill health, and we so step on the scale, choose the diet soda, go to the gym for a workout.

Participation in these little normalizations helps us to construct an architecture of certainty in what would otherwise be a very unpredictable social world. Thanks to the regularity of these sorts of normalizing rules and acts, we can be fairly certain that the person who dresses, looks, and smells like a man will have male anatomy under his clothes; that the charming professional colleague we have met only by phone will be, when encountered in person, wearing clothes and smelling good; that a newly discovered female model, when she appears in a swimsuit calendar, will have conventionally sexy legs—thin, smooth, hairless, without a sign of manliness.

Nevertheless, some people are born with anatomies that don't fit the social rules so far as anatomy and identity are concerned and that cannot be "normalized" through any simple procedure like shaving or the donning of eyeglasses. These people are born with anatomies that complicate efforts to easily categorize them. Cheryl Chase, for example, was born with mixed sex anatomy, internally and externally, which made it hard for people to figure out whether to expect her to become a boy or a girl. Lori and Reba Schappell were born conjoined at the head, an anatomy that can make a new acquaintance unsure whether they are to be approached as one person or two. Ruta Sharangpani is profoundly nearsighted but can see just enough to manage without an obvious aid like a cane or dog; she also has an eye that can't quite meet yours because it shakes and wavers. Danny Black has achondroplasia, a form of dwarfism, and though he is middle-aged he inhabits a body whose proportions are supposed to characterize only the immature.

Despite the fact that these people did not choose to have these bodies—at least not in any simple sense of choice—they are often treated as if they have intentionally violated a social norm, which in a way they have.[6] People avoid meeting their eyes, whisper about them, and act in a way that signals shame.[7] I find myself doing this. When, without warning, I encounter someone whose legs are bent so that he walks very strangely, or someone who is an unusual size for her age, I find myself feeling very awkward, making apologies for them or for me, struggling to get beyond my discomposure yet chewing on the image for a long time afterward. Most of us are so used to dealing with people who fit invisibly into the standard categories of anatomy and identity that it is jarring when we meet someone who doesn't. And it is the recognition of this awkwardness, the recognition of how comfortable it can be to be considered normal, how uncomfortable it can be to be considered abnormal, that motivates adults to want to surgically normalize children born with unusual anatomies, to separate the Loris and Rebas, to make the Cheryls look like "real" girls, to stretch the limbs of the Dannys, to make the Rutas look fully sighted.

Often the adults who impose such a normalization understand it as a charitable manifestation of pity. And no doubt it is. But "pity" is defined as sorrow for another's suffering or misfortune, and that's exactly why it is experienced by many people born with unusual anatomies as not only unsupportive but actively oppressive; for pity implies that the subject must be suffering and unfortunate. When I asked Lori Schappell how she felt when people treated her with pity, she bristled, saying that as soon as she saw such a "pity conversation" starting, she would end it or leave it.[8] Trying to fight the degradation of pity, Ruta Sharangpani told me once, is "like trying to climb a glass wall. There are no handholds, no way to talk to a pitying person, because she or he does not see the disabled person as a competent individual."[9] So, however unintentionally, pity silences the person who might otherwise speak to defend the value of her person and her life. That's why parents of children born with unusual anatomies often also insist they don't want or need anyone's pity. Patty Hensel, mother of the conjoined girls Abigail and

Brittany, told *Life* Magazine in 1996: "People say, 'We pray for you and the girls.' . . . But we don't need anyone to feel sorry for us."[10]

Yet at least until quite recently, sorrow and pity formed the narrative backbone of the usual story told about children born with unusual anatomies. The story went like this: These innocent, pitiful children are born cursed with tragic deformities; but through the miracle of modern medicine, doctors can remove the curse, changing them into normal little kids and saving them from a life of shame and mockery. Delivered into the happy realm of those who were created normal, they are henceforth free to live a full life otherwise unavailable to them.

But this book seeks to tell a different story. By considering conjoined twinning, arguably the most extraordinary form of human anatomy, in relation to other anatomical states that challenge cultural norms of identity—intersex, dwarfism, giantism, cleft lip (once known as harelip)—this book explores the extent to which anatomies do or must limit political and social identity, the extent to which a "deformed" or "malformed" anatomy must be pitiful. By considering conjoined twinning and other "deformities" within the larger historical context of anatomical politics, it argues for a more radical understanding of "abnormal" bodies. It seeks to change assumptions made about people born with unusual anatomies, and by doing so it seeks to change the context built around those people. The typical story told about such individuals is one in which the child's anatomy is changed to fit the social context. This book seeks instead to change the social context by exposing the breadth and depth of that context. It endeavors to show what something as rare as conjoinment could have to do with the rest of us.

To what extent are people who are conjoined abnormal? This is the subject of Chapter 1. There's no question that statistically they are extremely rare, accounting for perhaps as few as one in 200,000 births and no more than one in 50,000. But the reason they are treated so differently from others is not simply that they are rare; it is that people in general expect, quite reasonably, that any individual they meet will be the only person inhabiting his or her skin. Because most singletons—by which I mean people born with no anatomical bond to anyone but their mothers—understand psychosocial individuality as *requiring* anatomical individuality, they tend to assume that conjoined twins are trapped in such a way that makes a happy, normal life impossible. Only surgical separation could truly make them free.

New York Magazine vividly exemplified this assumption when it printed a photograph of infant twins Carmen and Rosa Taveras in November 1993—several months after they'd been surgically separated—under the headline "FREE AT LAST."[11] But as Chapter 1 demonstrates, such a headline would make little sense to people who are conjoined, because most people who are conjoined do not feel physically entrapped. They do not wish they had been born into singleton bodies. Indeed, Laleh and Ladan Bijani, who chose to be separated in 2003 at the age of twenty-nine, were the first conjoined twins in history to consent to separation surgery. Though it may seem shocking, in none of the hundreds of previous separation operations performed were surgeons given permission by the patients themselves to do the surgery.[12] This is not simply because most conjoined twins fear the risks of separation. It is because, as Chapter 1 elaborates, people who are conjoined typically feel that their bodies and lives are perfectly normal and acceptable—sometimes even preferable. They don't think there is anything fundamentally wrong with being conjoined. Thus, one of the ways in which conjoined twins are like almost everyone else is that they tend to readily accept, and even prefer, the anatomy with which they were born. So in light of what such people themselves have said, we might well understand conjoinment as an integral part of their individuality, paradoxical as that sounds.

Why, then, do doctors ever separate conjoined twins? Chapters 2 and 3 examine this question. Chapter 2 explores the types of surgery, including separation surgeries, which are explicitly aimed at physically normalizing children born with socially problematic identities. Of course there are many other ways in which we seek to normalize children (for example, through formal and informal education), often with beneficent intentions. But one must ask whether normalization surgeries really work in the sense of providing otherwise unavailable psychosocial health, whether they are the best route to psychosocial health and social justice, and what basic criteria ought to be fulfilled by any normalizing surgical procedure. While by no means arguing against all normalizing surgeries, I would like to problematize a process that is too often portrayed as a technological fairytale in which everyone but the dragon called Deformity ends up happily ever after.

The third chapter explores an extraordinary form of separation surgery in which doctors intentionally "sacrifice" one conjoined twin in an attempt to save the other's life. Increasingly common and increasingly controversial, these surgeries represent the only case in which surgeons are given explicit permission to separate a brain-live person (or at least her mind) from the organs keeping her alive, so that someone else may survive. Because such surgeries are motivated by the belief that both children will die imminently unless there is an intervention, and because surgeons are forced to terminate the life of one child to try to save the other, sacrifice surgeries are ethical hornets' nests and wrenching affairs for all involved. I'll discuss the history of sacrifice surgeries, focusing on three particular cases and the moral dilemmas involved in each.

Chapter 4 moves this story out of the surgical theater and back to the social theater, in part to consider the cultural context in which parents and doctors make decisions about normalizing surgeries. It looks at the way people with unusual anatomies have been presented, displayed, and exhibited over the centuries, by themselves and others, particularly in the medical sphere. Again departing from the usual story ("These poor people used to be stuck in freak shows, and now they're saved by medicine"), I pose three questions: What, exactly, was wrong with freak shows? Are we really past them, or have they just been shifted into legitimate forms by being medicalized? And finally, shouldn't we wonder why it is socially acceptable for some people—models, anchormen, basketball stars—and not others to make money from their unusual bodily differences? We'll look at some radical representations of people with unusual anatomies, and ask whether we should immediately condemn public displays of the odder forms of human anatomy.

There was a time when people with unusual anatomies appeared more regularly as the stars of live performances and film. As some readers will know, this book takes its title from Tod Browning's 1932 film *Freaks*—the story of a circus sideshow troupe that includes a pair of conjoined twins, played by the actually conjoined Daisy and Violet Hilton. In the film, a beautiful normal-bodied woman infiltrates the troupe and pretends to accept them, all for her own personal gain. They, in turn, accept her, chanting: "One of us, one of us!" The chant recurs at the conclusion of the film. By then, the "freaks," resentful of the woman's now-obvious contempt for them, arrange to have her turned into a sideshow exhibit. Once her transformation is complete—once she is normalized into their world—they repeat their call: "One of us, one of us!"

It's a pretty gruesome ending. Nevertheless, I chose the phrase "one of us" for this book's title partly because it captures the paradox of accepting conjoined people as *one* of us, and partly because I think the scenario in the film plays out the fundamental anxiety elicited by people with profoundly unusual anatomies, including conjoined twins: they expose the socially tenuous nature of all human anatomies and raise the question of who ought to count as normal or ideal. So this book ends not with a shot of horror, as *Freaks* does, or with the suggestion that all "normates" be made abnormal.[13] Rather, it concludes with a question about the future of anatomy and its relation to social and political identity.

The great theorist of racism W. E. B. DuBois asserted that the problem of the twentieth century was the problem of the color line.[14] I argue in Chapter 5 that we are far from having solved this problem, and that the problem of the twenty-first century will be the great fault on which the color line falls: the anatomy-identity line. It's true that we've come a long way since the days when power and wealth were viewed as a birthright, men were considered inherently superior to women, and whites were seen as inherently superior to all other races. Scientists are continually revealing cracks and holes in anatomical borders once thought to be seamless: female/male, black/white, human/animal, even living/non-living/dead. The progress of science and of democracy continues to decouple social and political identity from anatomy. And an underground "freak chic" combined with an aboveground disability rights movement is slowly eroding the hegemony of the normate. Yet all the while we see attempts to reestablish clear categories, norms, and authorities—attempts that include the FDA's approval of Botox as a medically administered antiwrinkle treatment, the Supreme Court's efforts to narrowly construe the category of "disabled," the scientific search for the "violence gene," and a bouncer's decision to keep a man out of a lesbian bar. So we will look at the ideals of Enlightenment progress, the tools of medicine and technology, and the knowledge of how anatomies become pleasurable or painful, and see whether these can be used to construct a rational, democratic civilization that grants full membership to people born with socially challenging anatomies—recognizing that human civilization as we know it cannot exist without anatomical norms. How is the social context of conjoinment and other unusual forms of anatomy likely to change in the coming years? Who, in the future, will count as one of us? And how will we indicate that belonging? Through a politically correct labeling process? A protected right? A guaranteed genetic endowment? An ensured normality?

People often ask me how I came to this topic, and I think that my journey is worth describing. It will clarify my point of view and—I hope—encourage readers who, like me, think of themselves as normal-bodied to consider the ways in which this book is about them too.

When I was earning my Ph.D. (in the history and philosophy of science), my graduate advisor suggested that I study the history of the biomedical treatment of hermaphroditism, since the topic of gender was one of my interests. This sounded good. I thought I would look at Darwin's work on hermaphroditic barnacles, the history of embryology—that sort of thing. My professors kept suggesting I look at human medicine. I couldn't understand why. I didn't think hermaphroditism—now usually called intersex—ever occurred in humans past the embryonic stage. But eventually I bowed to pressure, looked at the medical literature, and was astonished to discover thousands of documented cases of people with unusual sexual anatomies. Some of them had apparent female anatomy but no vagina. Some had apparent male anatomy, except that their penises were very small. Some had ovotestes, gonads that were part ovarian and part testicular. Some had genitals that seemed to be halfway between a boy's and a girl's.

I pursued this research and started to publish history articles about what happened to people with these intersex conditions in the late nineteenth century, before the age of surgical "correction." Soon after, people who had been born with intersex conditions began learning of my work. Some of them looked up my email address and contacted me to talk about their experiences. One of those people was Cheryl Chase, founder of the Intersex Society of North America, a policy and advocacy organization whose stated mission is "to build a world free of shame, secrecy, and unwanted genital surgeries for people born with atypical reproductive anatomies."[15] Cheryl considered herself a victim of an unethical and scientifically unfounded system of treatment that takes children with intersex conditions and changes their anatomies, through surgery, hormone treatments, and shaming silence, to try to conceal any sign of intersex.[16] The goal of these treatments is to make intersex children look and act like "normal" boys and girls. This system is motivated by the belief that no child with intersexed anatomy could grow up to be normal, healthy, and happy, and that it would be cruel to leave the child's "ambiguous" anatomy intact. But soon after Cheryl and I started communicating, she told me that she and many other people with intersex suffered from difficulties stemming from surgical interventions: limited sexual sensation, iatrogenic medical problems, odd-looking surgically constructed anatomy, estrangement from their families, and an overwhelming sense of shame and freakishness. In her view, the medical system was *causing* the very psychological, familial, and social harm it was supposed to prevent. And she asked me to help change the system.

I was a little embarrassed that my reply sounded like Dr. McCoy's disclaimer on *Star Trek:* "I'm a historian, not a doctor or sociologist or activist." But Cheryl was unrelenting, and finally, with her encouragement and with the help of my medical-student husband, I began to look at the present-day medical literature. I suppose I was expecting to find a generally safe, effective, modern system of treatment that perhaps left a few minor problems unaddressed. Instead, I found unproven claims that a girl born with a large clitoris (larger than $\frac{3}{8}$-inch when stretched at birth) would grow up a tomboy or a lesbian; that a boy born with a very small penis (smaller than 1 inch when stretched at birth) had to be castrated and made into a girl; that in cases of intersex doctors had an *ethical duty* to withhold personal records and facts from patients and parents (though this would be considered unethical in other contexts) because otherwise the normalizing treatment wouldn't work and the patient might commit suicide.[17]

Although this system obviously suffered from major ethical and empirical flaws and a fair degree of Victorian sexism, I thought that reforming it would be simple. Cheryl and I and our handful of allies could publish some articles and talk with a few of the big players in the medical field, and the system would then be revised to reflect modern ethical sensibilities and a modern understanding of human sexuality. From my historical research, I knew that the concealment treatment system was only about fifty years old—relatively new, at least to a historian—and that there was plenty of evidence showing that people with intersex conditions prior to the era of childhood normalizing surgeries did well psychologically and socially.[18] But nearly all of the physicians we talked with claimed that Cheryl was just an unfortunate bad outcome—that the system was basically sound. They said they were sure their patients were happy, mentioning the fact that one or two had invited them to their weddings. Moreover, they kept telling us that boys and girls with intersex ought not to grow up "uncorrected" because they would be exposed to comment in the locker room at school or the diaper-changing station at daycare. People would not be able to handle it; the kid would suffer needlessly. The only way to handle unusual anatomies was to "fix" them, to make them look more normal.

Part of me thought that perhaps they were right. I was incorrectly assuming that, prior to getting to know Cheryl, I had not met anyone with an intersex condition—that it must be pretty rare, or so damaging that people who were "uncorrected" were dead or in hiding. But Cheryl, bent on educating anyone willing to listen, started introducing me to people with intersex conditions, their parents, and their lovers, so that I could hear their stories and see that she was not the only one who had suffered under the current medical system. Cheryl and I began collecting these stories and publishing them, adding to the other stories that were appearing on the Web and elsewhere.[19] Again and again, I heard first hand (and also second hand from disillusioned doctors and nurses afraid to speak out against the establishment) about well-intentioned attempts at normalization that had left people feeling physically and psychologically damaged. I also met people who had slipped through the cracks and escaped the medical system for various reasons. Some were aided by extraordinary parents who had rejected medical advice; others had been too sickly as young children to withstand such treatments. All of these people told me that they saw their bodies as different but normal, and that they had no interest in "corrections." They *liked* their bodies. Some volunteered that they *preferred* their intersexed bodies to the norm.[20]

At that time, around 1996, a lot of Americans—feminists, doctors, legislators—were up in arms about "female genital mutilation" (an array of procedures sometimes also called "female circumcision"), practiced in certain regions of Africa. Though proponents of traditional female genital cutting were defending it as an important and worthwhile · cultural practice, most Americans considered it a barbaric system in which healthy genitals were being cut purely for societal reasons. To advocates of intersex reform it looked a lot like the treatment of intersex, except that female circumcision wasn't performed in a sterile hospital environment and tagged with an insurance billing code. Children were having their genitals made "normal" according to cultural definitions of normal. The practice could cause loss of genital sensation, incontinence, a host of related health problems, and sometimes even death, all in the service of upholding narrow ideas of what it means to be a girl, boy, woman, or man. Yet few of the people outraged by the African practice agreed that intersex was anything like this. Indeed, the 1996 federal legislation banning female genital cutting in the United States incorporated a special exemption designed to protect the standard medical treatment of intersex—a clause that in itself was an acknowledgment of the extent to which female genital cutting looked like intersex surgeries.[21]

Because Cheryl was having such a hard time persuading people to help her, she asked me to think about why I might have been more willing than others to take seriously the problems with the treatment of intersex. So I started to think about my personal history, wondering what in particular had aroused my sympathies. I realized that, like some of the people I met who had intersex conditions, I could be said to be suffering from a genetic sex condition.[22] Though I had not actually had my genome screened, all the anatomical signs of Double-X Syndrome were there. First there were the associated physiological disorders, including cyclical bouts of pain and bleeding, and hair growth patterns that differed from those of people born without Double-X. Then there were the psychosocial pitfalls: people with Double-X are more likely than others to live below the poverty line, are much more likely to be sexually assaulted, and by law are prohibited from marrying people with the same condition. Some prospective parents have even intentionally aborted fetuses diagnosed with Double-X, in an effort to avert the tragic limitations the syndrome often entails.

Double-X goes by a more familiar name: womanhood. For centuries, women in Europe and North America were seen by the medical establishment and many in society as biologically inferior, inherently ill, undeserving of rights to which others (namely, men) were entitled. Even today, girls and women disproportionately suffer from many forms of sexism and gender-based violence. Yet to improve the situation of women, sympathetic people didn't and don't ask surgeons to change women into men; they sought and seek to fix the oppressive social system. Why not do the same for girls, boys, men, and women with intersex conditions?

Then there was my brother Paul. My parents first brought Paul into our family as a foster child at four months. We all fell in love with him, and my parents formally adopted him as soon as they legally could. But he was always obviously different because of his anatomical condition: the rest of our family was white, our neighborhood was white, our schools were white, and my brother was black. Biologically he's actually multiracial, but in such a segregated place as my Long Island home-town, he was effectively black. And by virtue of being black in a place where to be black was to be abnormal, Paul might have counted as disabled according to the basic definition since provided by the Americans with Disabilities Act:[23] people regarded him as having an anatomical impairment that substantially limited his major activities. Some neighbors sought to limit the kinds of people he could date. Teachers assumed he wasn't capable of learning as easily as his white sisters and brother, and this limited his education and sense of self. People tended to stare at him, to treat him as if he carried a clear stigma, especially when as an adolescent he was seen with someone like me who was a "normal" (white) female.[24]

A couple years into my work with Cheryl, her views were challenged on *Dateline NBC* by a physician who was a specialist in intersex. The physician asked the camera rhetorically, "In our society, we're [so often] not accepting of somebody who's a different color than we are. . . . How are we going to accept somebody whose genitals are different?"[25] It was then that I finally decided to ask my brother how he felt about his anatomical difference, something I'd never had the courage to do before. Did he wish he had been born white? Did he wish he had been raised in a family and neighborhood where others looked like him, where he would have been "normal"? His answer was that he had always felt normal, and that he only wished he'd had someone else like him to talk to when he was growing up. No, he didn't want to be white, doesn't want to be white, even though he now has a clearer sense of the racism to which he has been subjected.

It's true that, growing up, my brother at least got to see people like him on TV—even if it was only Fat Albert and Bill Cosby—and that, until recently, children with intersex conditions or children who were conjoined never saw anyone at all like them on TV. (Zack and Wheezie, brother-and-sister conjoined dragons on the PBS animated children's series *Dragon Tales,* have changed that a bit.)[26] Nevertheless, there are clear similarities among the conditions of being black in a white world, a woman in a man's world, and conjoined in a singleton's world. In all of them, anatomical difference is assumed to be some sort of determiner of one's future: different body, different person, different life. These assumptions then become self-fulfilling prophecies. In all of these cases, difference tends to be seen as a cause of suffering, and only suffering—something to be pitied. Most normates assume that everyone, given the choice, would choose a "normal" body. But many of the "fixed" say they feel damaged by being "fixed"; and many of the "unfixed" claim that they feel normal, that the people who know them well—as individuals, not simply as examples of various conditions—see them as normal. All this is true of people who were born missing limbs, people who were born blind, people who were born black or female or with an intersex condition. It surely isn't a coincidence that historically there have been some close ties between the civil rights movement and the disability rights movement.[27] So why, I wondered, do we try to change the children in some of these cases and change the world in others? Why not change minds instead of bodies?

Let me make clear that I am not interested in romanticizing conjoinment or dwarfism or intersex, or any other anatomical condition. What I am attempting to do here, with stories, history, and analysis, is to suggest that there is another way to think about "deformity" other than as a medical tragedy that needs fixing. In fact, most atypical anatomies—conjoinment, intersex, cleft lip, and so on—are, by this point in history, fairly well documented human experiences. And in that documentation we discover that they are remarkably like many other human experiences, such as motherhood, marriage, "racial" differences, and radical size variations—similar enough that we ought to reconsider our views of people with unusual anatomies. We should take seriously the possibility that they are entitled to what the rest of us are, including a validated sense of normality and a reasonably wide degree of self-determination. Right now, they don't have that. I'm thinking: enough pity already.

The Limits of Individuality

When I met Tara and Sarah, two years before I started writing this book, it was striking how much smaller Tara was than Sarah. Their size difference was quite dramatic. So was the difference in their intelligence. Sarah, the larger, was much more physically advanced than Tara in terms of motor development and much further along in mental development, though Tara gave indications that she'd be able to progress. The two of them obviously loved each other very much and took great pleasure in their physical closeness. When I talked to Sarah about her situation I learned that, along with their anatomical intimacy, a point of consternation for some people was the fact that Sarah was essentially keeping Tara alive. Every day, through their anatomical union, Sarah provided Tara with all of the nutrients Tara required. In this way their relationship was quite unequal; Tara simply couldn't do the same for Sarah. But Sarah wasn't bothered by this and felt it was just a natural part of who they were; it wasn't really anyone else's business. Nevertheless, about four months after I'd met them, Sarah decided—partly in response to societal pressures, partly in response to her own growing need to be separated from Tara—to pursue a medical intervention that would allow them to become physically independent. Henceforth, Tara would have to depend on artificial means of feeding.

I never had the chance to meet Charlie and Edward, identical twins who died long before I was born. By all accounts—their own writings, stories about them, and oral histories passed down through their numerous descendants—they had quite distinguishable personalities. Throughout their lives they remained close and enjoyed each other's company; even as old men, they couldn't recall ever having fought with each other more than twice. They loved to travel, and because of the privilege afforded them by their anatomies, they were able to leave their tiny hometown and trek around the world together, meeting many interesting and famous people along the way. When they were in their early thirties, Charlie fell in love with a young woman named Adelaide. Soon after, Edward fell in love with Adelaide's sister, Sallie. Each married his sweetheart. Years later, when they could afford two residences and when their wives expressed a preference for this, the couples set up separate households and farms about a mile apart in Mouth Airy, North Carolina, so the brothers' families could stay close but not too close. Charlie and Adelaide eventually had ten children; Edward and Sallie, twelve. The brothers were able to support their large families in part by owning slaves, another sign of the sort of privilege they enjoyed as a result of their particular anatomies. In 1874 Charlie and Edward died a few hours apart, at the age of sixty-two.

If you know the histories of Chang and Eng Bunker, the first conjoined siblings ever referred to as Siamese Twins,[1] you'll have realized that I changed their names in the last story to Charlie and Edward. I did so not to discount the importance of Chang and Eng's background— they were of Chinese heritage and came from Siam—but rather to defamiliarize their story a bit, to point out the ways in which Chang and Eng carried on separate lives while being conjoined and to offer a perspective different from that of the classic story singletons tell about people who are conjoined. Usually singletons speak of conjoinment as a fleshy prison that *limits* individuality and freedom. But Chang's and Eng's stories, when considered apart from the typical singleton anxiety, read as the stories of two men whose conjoinment actually may have *accentuated* each man's individuality and *increased* many of his freedoms—for example, the freedom to travel the world and to earn great sums of money through performance, money that opened still more doors to them. Ironically, far from enslaving them, conjoinment brought Chang and Eng such affluence that they could afford to enslave others for their own economic benefit.

What, then, of Sarah and Tara? Their story is a fictional compilation of the stories of nursing mothers and babies I have known. Sarah is the mother; she is keeping Tara alive by breastfeeding. The "medical intervention" chosen by Sarah—the intervention that allows them to be separated—involves the use of silicone nipples, plastic bottles, and infant formula.

Now, I'm not going to claim that being conjoined is just like being a nursing mother or baby. Hardly. Instead, I opened with this comparison to suggest the value of putting aside what seems perpetually obvious—how different being conjoined must be from being a singleton—to think about the less obvious: how being conjoined is *like* other human experiences. Such a move can be valuable in two ways. First, thinking about the way in which conjoinment is like other states helps us parse out the reasons conjoinment is so socially troubling (for example, like breastfeeding, it makes public a physical intimacy that usually marks only private relations). Second, considering how conjoinment is like other experiences opens a window of sympathy that may allow singletons to take seriously the positive claims people who are conjoined make about themselves and their physical situations.

Eng and Chang were born in Siam (now Thailand) in 1811. They were healthy boys joined together by a flexible round band of flesh stretching roughly from sternum to umbilicus, a band about five inches long and about nine inches around at its thickest (see Figure 1).[2] Their mother, undoubtedly surprised by her offspring's anatomy, nevertheless raised them pretty much as she did her other children. As the modern-day biographers Irving Wallace and Amy Wallace have noted, "Their mother Nok was determined that they should grow up as normally as possible, and she neither ignored the twins nor pitied and overprotected them. She was matter-of-fact about them."[3] They, in turn, gave her every reason to see them as typical boys. They grew strong and agile, played rambunctiously, learned quickly, and helped from an early age to support

1. Chang and Eng Bunker at the age of twenty-eight, in a lithograph from 1839. (The artist accidentally drew the twins in reverse; Eng would actually have been on the left if you were facing them.) Chang is holding a booklet sold as part of their tour.

the family, first as fishermen, like their father, then as merchants. Word of their unusual anatomies spread—they were widely known as the Chinese Twins in Siam, because of their Chinese parentage—and they received countless invitations from people eager to see them, including the king of Siam and, later, citizens of far-flung countries. In 1829, during the great age of unabashed exhibition of anatomical curiosities, the teenagers were taken to New York by "an enterprising American ship-captain . . . for the purposes of exhibiting them."[4] Their mother received a small monetary compensation for the loss of her dear and helpful sons, whom she would never see again.[5] For their part, the boys, strong and intellectually vibrant, were thrilled at the chance to travel the world.

As young men, Chang and Eng took the surname Bunker while touring the United States. Their tours paid pretty well; by their early thirties they had made enough money to buy some fertile land and a few enslaved persons in North Carolina and to settle into a quiet life of farming. Intelligent, outgoing, hardworking, and well read, Chang and Eng were widely respected in their new hometown. In April 1843, Chang married Adelaide Yates and Eng married her sister, Sallie, "daughters of an American clergyman."[6] The wedding took place despite reservations on the part of the brides' family. Judge Jesse Franklin Graves, a good friend of Eng and Chang, later recalled that the parents' "objection to these gentlemen did not arise from any want of character or social position, for in point of morality, probity, strict integrity, they sustained a spotless reputation, but it had its origin in an ineradicable prejudice against their race and nationality."[7] Perhaps Adelaide and Sallie were able to see beyond the men's race (and conjoinment) because the sisters had long been familiar with anatomical stigma: their mother was extraordinarily fat, and widely known for this.[8] Nevertheless, as Judge Graves noted, Chang and Eng were also good catches—well off, personable, devoted, financially savvy, and highly esteemed by those who knew them.

Apparently following the wishes of their wives, who found living as one big and growing family increasingly difficult, Chang and Eng set up separate households on nearby farms in 1857.[9] Eng and Sallie eventually had a total twelve children, while Chang and Adelaide had ten. Toward the end of their lives, the Civil War deprived the Bunkers of their slaves and this made finances tight, so the brothers resumed their old occupation. They made a grand tour which included visits to the United Kingdom and Europe, as well as appearances all over the United States (see Figure 2). The family letters from this period were donated by a descendant to the archives of the University of North Carolina at Chapel Hill.[10] They reveal the twins as being both practical and tender, mulling over questions of when to sell the crops at market, inquiring after the health of individual family members and neighbors, worrying when family news was late in coming.

In 1870, on the ship bringing them home from Europe, Chang suffered a stroke which resulted in paralysis on his right side. After their deaths, the newspapers played up how hard it must have been for Eng to bear up with an attached yet disabled brother, but Judge Graves remembered Eng as having been warmly supportive, in keeping with the loving relationship the two had enjoyed throughout their lives: "Eng's treatment of his brother was very kind and forbearing during all the long period of his sickness, showing great tenderness and affection for him and endeavoring by every means in his power to alleviate his suffering. His kindness was received with the warmest appreciation by Chang, whose disposition was very different from the morose, ill nature so falsely ascribed to him"[11] by sensational press reports which sought to enliven the drama of the men's conjoinment.

Though Chang and Eng coined the term "Siamese twins" to refer only to themselves, the moniker continued to be used long after for all conjoined twins, at least colloquially. Today, people who are conjoined and those who study them would like to see the misleading phrase "Siamese twins" abandoned in favor of "conjoined twins." Yet in many ways the Bunkers' case is paradigmatic beyond the name. The Bunkers, like all dual-consciousness conjoined twins, developed distinct but cooperative personalities. They occasionally showed themselves to others for financial gain, but they thought of themselves primarily as something other than performers—as family men, as farmers, and so on.

2. Eng and Chang Bunker in their later years, shown with Eng's son Patrick Henry on the left and Chang's son Albert on the right. The drawing was made during their grand tour in the late 1860s.

And like many conjoined twins who survive childhood, Chang and Eng were educated, generally healthy, and enjoyed the company of friends and neighbors, who apparently treated them pretty much as they treated singletons they knew.

The Bunkers' case is also paradigmatic in terms of the way it was handled by the medical and scientific establishment. Medical and scientific experts of the nineteenth century tended to feel they had a special claim to the Bunkers' bodies, and the question they wanted answered most, the *questio vexata,* was this: "Could the twins have been safely separated during their lifetime?"[12] This question had not much vexed Chang and Eng themselves; apparently, they seriously considered separation only once, late in life, and even then they seem to have done so only because their wives wanted it.[13] (They occasionally did talk with doctors about separation purely to drum up profitable publicity.)[14] But medical and scientific experts were intently focused on the question of separation every time they saw or wrote of Chang and Eng. And they made it clear that, regardless of whether the brothers' well-being would have been improved by separation, the doctors believed separation of all conjoined twins was necessary for the well-being of the *social* body. In the words of the man who performed the autopsy on Chang and Eng, their condition as sexually active conjoined twins "shocked the moral sense of the community."[15] Indeed, so problematic were the Bunkers' sexual biographies to those who did not know them personally that medical experts took it as the "duty of the medical profession . . . to make an effort to elucidate the point at issue" (namely, the possibility of surgical separation), in order to inform future cases—in order to try to avoid more Chang and Eng Bunkers. The anatomists also considered it "a duty to science and humanity that the family of the deceased should permit an autopsy" to answer the *questio vexata.*[16] So two weeks after the brothers' deaths, learned men from Philadelphia came to fetch Chang and Eng's remains from the Widows Bunker. Sallie and Adelaide had been keeping the remains in the cellar in a tin coffin under charcoal, apparently in a sensible attempt to foil grave robbers.[17] The bodies would have been worth quite a bit, as the family knew from incoming letters offering large sums,[18] but Sallie and Adelaide chose to release them to the Philadelphia doctors without compensation.

On Wednesday, February 18, 1874, in the hall of Philadelphia's College of Physicians, a rapt audience of professionals heard the report of the autopsy performed on Eng and Chang Bunker. The internal examination had revealed that the connecting band of flesh consisted of skin, blood vessels, cartilage, and a bit of liver.[19] On the question of whether this finding meant that the twins could have safely been separated during their lifetime, experts largely concluded in the negative. Most decided that since the peritoneum (body cavity) extended into the band, the twins would probably have died from infection if separated. (Today, because of advanced imaging and surgical techniques, separation of this kind of conjoinment would be fairly simple and relatively low-risk.) Before releasing the brothers' remains to the anatomists, the families had requested that the bodies never be fully cut apart (this stricture was apparently stated in Chang's and Eng's wills),[20] and the anatomists respected that wish. But to judge from existing records, the wives did not realize the College of Physicians would embalm the bodies, delaying natural decay, or that they would decide to keep some of the internal organs, including the conjoined liver. (This liver, as well as a plaster death cast made of the brothers, is now on public display at the college's Mütter Museum.)[21] The remains that the family managed to recover were interred on Bunker land. Later they were moved to a joint gravesite behind the Baptist church in White Plains, North Carolina, a small white structure which Eng and Chang—strong, agile men at the time—had personally helped their neighbors to erect.

In July 2002 I visited the brothers' gravesite for the first time. I had been invited to North Carolina for their descendants' annual reunion, held across the street from the churchyard. Several hundred of Chang's and Eng's progeny were in attendance. One descendant of Eng and Sallie Bunker displayed her pastel rendition of the house she grew up in, the house where Eng and Sallie had once lived with their large family. The matriarch of the clan, Jessie Bunker Bryant, great-granddaughter of Eng and Sallie, presented her just-completed book of family genealogy, a magnum opus tracing the births, deaths, marriages, and accomplishments of all the descendants she'd been able to trace.[22] When I was invited to the microphone, I took the opportunity to encourage family members to visit Chapel Hill and see their ancestors' correspondence and account books. Many in the family were unaware of that remarkable collection.

The evening before the main reunion luncheon I had dinner with Margo Miles-Carney, great-great-granddaughter of Eng and Sallie Bunker, in the nearby town of Mt. Airy. Margo described how she'd learned of her famous ancestry, the night before her wedding. After the rehearsal dinner was over and the couple and Margo's mother had returned to Margo's apartment, her mother told Margo she had to speak to her in another room. Upon getting Margo alone, she had suddenly announced in distraught tones, "You can't marry Jack!" She then went on to tell Margo that they were descendants of Eng Bunker, one of the "Siamese Twins." Margo, unfamiliar with the story of Chang and Eng, went back to her fiancé and told him what her mother had said. An amateur historian well-versed in U.S. history, Jack was thrilled and went ahead with the wedding even more enthused. Margo's mother, however, continued to be deeply troubled about the family secret.

When I asked Margo why her mother had been so afraid and upset, she said she'd gotten the impression that Eng's conjoinment was not the problem. She thought it might well have been his race. Her mother had been worried that Margo might someday give birth to a child who looked Chinese. Indeed, Margo remembers her mother reminiscing anxiously that when Margo was born, she had a triangle at the base of her spine which her mother took as the sign of her Chinese lineage. She hid her recognition of it from Margo's father, and apparently could never bring herself to talk with Margo's father about any of it.

The first time Margo went to a Bunker family reunion and described how she'd learned about her extended family, another relative—a woman about her mother's age—had comforted her, saying, "Honey, nobody talked about it." This relative, too, had been informed in an abrupt way by her own mother (in her case on the occasion of her first period), as a warning about the family's Chinese blood. She agreed with Margo's suspicion: that their parents, like those of Sallie and Adelaide, had felt shame and fear not because Chang and Eng had been conjoined but because the family was of mixed race.[23] Though the stigma of conjoinment was unlikely to reappear in the family, the Bunker descendants, generation after generation, continued to fear the stigma that Americans assigned to people who were "interracial." Even the genealogist Jessie Bunker Bryant, who is intensely proud of her family history, came to realize that "the Siamese were dark-skinned people, and I guess back then marrying another race was a lot less accepted than it is today."[24]

As Margo Miles-Carney worked to uncover her family history, she came to learn about the lengthy history of oppression—informal and institutionalized—that Asians and Asian-Americans have experienced in the United States. She now knows that, throughout the nineteenth century and even in the twentieth, Chinese people were sometimes displayed in sideshows and museums just because they were Chinese, and that her mother's fears about being identified as Chinese would have been particularly intense during the Second World War, when people who looked "oriental" were treated as suspect.[25] The Bunker family has long worked to avoid and overcome racist assumptions. At the 2002 Bunker reunion, someone showed me a picture of a sign posted in a Mt. Airy convenience store during the previous year's reunion:

WE HONOR ENG AND CHANG BUNKER HERITAGE AND LEGACY
5 GENERATIONS OF INTEGRITY AND SERVICE

As in Chang and Eng's own time, the family has worked to be known and appreciated locally for integrity and service—for the content of their characters, more than for the color or shape of their skin.

Archeology and biology give us reason to believe that conjoined twinning has been a feature of human life since time immemorial. Statues and carvings from early civilizations seem to represent people born conjoined, and conjoinment has been observed in dozens if not hundreds of other species—cats, dogs, snakes, and so on. So it has probably

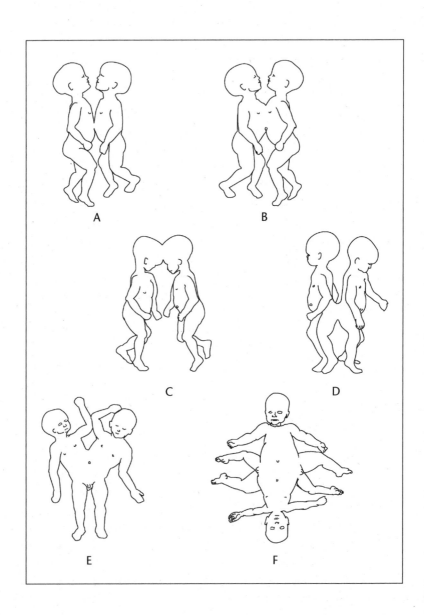

A

B

C

D

E

F

always occurred with some regularity, though prior to modern recordkeeping and literacy, it must have seemed even more extraordinary than it does today.

Most modern theorists believe that conjoinment occurs when an embryo begins to divide into two separate, identical embryos but does not complete the division.[26] So far as we know, every pair of conjoined twins, no matter how different they look from each other (and some look *very* different), are genetically identical. The medical names given to the different types of conjoined twins focus on the point of conjoinment (see Figure 3). Parapagus, or those intimately joined at the pelvic region (and sometimes also much farther up the body, toward the head) account for roughly a quarter of reported cases of conjoinment; in parapagus conjoinment, it generally looks as if there are two persons at the upper end of the body and one at the lower end. The reverse, where it looks as if there is one person at the head but two at the bottom, is called cephalopagus, and this type accounts for about 11 percent of cases reported. Thoracopagus twins, or those joined along the chest, account for about 17 percent; omphalopagus, or those joined at the umbilical region, account for about 14 percent; and ischiopagus, or those joined at the hip, account for about 12 percent. The rarest types include craniopagus (joined at the head; about 4 percent), pygopagus (joined at the lower vertebrae and buttocks; about 4 percent), and rachipagus (joined at the spine, back to back; less than 2 percent).[27] The causes of conjoinment aren't clear—that is, no one knows why some identical twins fail to separate completely—though experts have abandoned the centuries-old belief that a mother's thoughts, desires, anxieties, or dreams might be precipitating factors. Scientists now think

◀ 3. Some types of conjoinment. The medical custom is to label conjoined twins according to the location of the conjoinment, so the twins shown here would be labeled: (a) omphalopagus; (b) thoracopagus; (c) craniopagus; (d) pygopagus; (e) parapagus; (f) ischiopagus. There are many other types, and conjoined twins who are classified by these terms do not necessarily look exactly like the versions shown here. For example, craniopagus twins may be joined at the top of the head rather than near the forehead. Dramatic asymmetry is also possible.

that, as with many developmental anomalies, environment and chance both play a role.

Conjoinment doesn't always involve two fairly equal and healthy bodies, as it did in Chang and Eng's case. Occasionally, one twin develops fully while the other fails to make it past the embryonic stage, so that a person might go through life without knowing she has a tiny, undeveloped twin attached to some point of her body.[28] Sometimes one twin develops fully while the other develops only a few body parts, such as legs or arms. These "incomplete" pairs are known in the medical literature as "parasitic" twins (see Figure 4).

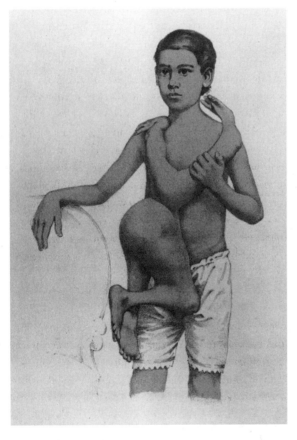

4.
A boy named Laloo, who was born in the late nineteenth century with a "parasitic" conjoined twin.

Estimates of the incidence of conjoined twinning in humans vary from 1 in 25,000 births to 1 in 200,000. Reporting has not been stable enough to yield a reliable estimate, and it is unclear whether some instances have been counted as one birth or two. In any case, we can be fairly confident about a few general facts: the phenomenon is unusual but occurs with some regularity; roughly two-thirds of conjoined twins are female;[29] 40 percent or more of conjoined twins are stillborn; another 35 percent die within one day as a result of profound medical problems stemming from their anomalous development.[30] (More often than not, conjoinment comes with major medical problems caused by organ sharing, incomplete development of physiologically important parts, or problematic circulatory patterns.)

Perhaps it is because viable conjoined twins are so rare that those of us who are singletons have such a hard time imagining what it would be like. One modern-day author, while noting that "Chang and Eng seemed to have been a success" at day-to-day life, remarked that "two people never being able to obtain privacy to bathe, excrete, copulate or eat defies imagination."[31] Yet a close reading of the many biographies of conjoined people clearly shows that their lives are not necessarily horrible, unbearable, or even that unusual. Some pairs have lived reasonably long lives, as the Bunkers did. Quite a number have had lovers, and a few, again like Chang and Eng, have had children and families. Many have traveled widely, been well educated, enjoyed occupations. Some have had positively boring lives worthy of the most "normal" of us. Indeed, more than one student of the phenomenon has concluded that "even nonseparated twins can go on to lead reasonably normal lives."[32] Whether they can, of course, depends to a great extent on the social situation around them.

In the United States, conjoinment might be especially challenging because American culture equates individualism with independence, and interdependence with weakness. In the United States, it seems as if only children can be physically dependent on others without risk of shame, without risk of being seen as enjoying less than full personhood. (And even in that case, cultural norms limit the extent to which children can be physically dependent without risk of shame; many people frown on the idea of children sharing their parents' bed or nursing beyond early infancy.)[33] The "right" to individuality represents one of the most pervasive American mores. If someone objects to an American's decision to buy an SUV or a "breast job"—purchases that could be seen as wasteful and selfish uses of common resources—the typical (and typically successful) defense is an assertion of the buyer's right to individuality.[34] To be true to yourself as an American, you must show yourself to be different, separate, distinguishable from all others. Being an individual in the United States does *not* mean being an integrated member of a community, as it does in some cultures—cultures where conjoinment might be easier to live with for precisely this reason.[35]

Indeed, American mythology is largely a celebration of the independent person. Consider how stories of important inventions like the sewing machine and discoveries like penicillin—inventions and discoveries which really took generations of development—are credited in schoolbooks to one man or one woman. Not uncommonly, a key element in such stories is the individual's struggle against the system (i.e., other people) in an effort to rise above the fray. Even the American geographic imagination seems to be largely about equating freedom with expansive physical space. I'm reminded of the complaint often attributed to Abraham Lincoln's father: the first time he stood on his doorstep and saw smoke rising from a nearby chimney, he opined that the neighborhood was getting too crowded.

But even within this context—or perhaps as a *result* of living in this context—Americans who happen to be conjoined feel that they, too, are independent individuals. They do not think they need a discrete body to achieve independent status, any more than another person needs to grow her own food, sew her own clothes, and be her own doctor to count as an independent individual. The fact is that across cultures and across time, the great majority of people who are conjoined simply have not expressed the sensation of being overly confined, horribly dependent, physically trapped, or unwillingly chained to others. In my own experience, the people who complain most of these sensations are present-day Californians living in the Midwest.

Ruthie and Verena Cady were born in Durango, Colorado, on April 13, 1984. Their mother, Marlene, remembers that after the birth everyone in the delivery room fell silent. When she asked what was going on, her doctor answered: "You have two beautiful little girls. They are very special. They are attached."[36] Marlene's "head felt numb, and the words didn't register. I thought, 'Oh, well, okay, let's just pull them apart. Just unsnap them or unzip them or whatever you do.'"[37] But it wasn't that simple. Ruthie and Verena were joined from breastbone to waist, sharing one weak, three-chambered heart and some intestinal tissue.

After evaluating the twins, doctors advised Marlene and her husband, Peter, that separation "would mean the certain death of one twin and a 90 percent chance of death for the other."[38] Deciding against such a risk, the parents took the girls home and, in view of the bleak medical prognosis, expected them to die within a year. But Verena and Ruthie lived to the age of seven, defying all expectations. Not only did they learn to walk (with the persistent help of their mother and a physical therapist); they eventually took to dancing, riding a special tricycle, and roller-skating. They went to kindergarten and first grade at public school and played regularly with their sister and other children, often using supplemental oxygen to compensate for their increasingly poor respiratory health.

The day Marlene took her conjoined babies home from the hospital, a man standing outside the building shouted to her, "Hey, lady! Are those the Siamese twins? Can I see them?" Marlene "cringed. But the young man who approached was not at all malicious—simply curious. He asked plenty of questions, but they were compassionate. The caring in his voice was a revelation."[39] And that was a sign of things to come. After their arrival home, the Cadys found their neighbors extremely accepting and helpful—an experience that was repeated when the family moved cross-country to Cranston, Rhode Island. The change to a lower elevation also alleviated the girls' breathing problems. At times, Marlene and Peter felt overwhelmed by the medical needs of their twins, but their strong Christian faith served them well during crises. "Maybe the situation was created so that we could grow closer to God if that was our desire,"[40] Marlene speculated. "There are people who come up to us and say, 'Oh, how tragic, how tragic.' I always tell those people that the only tragedy is in their interpretation of the girls' situation, because obviously Ruthie and Verena are happy kids. We allow them to explore, and they find their own limitations. We don't set limits for them."[41] Marlene maintained an admirable philosophy of parenting: "As long as their environment isn't saying 'no,' but 'yes, you can,' they'll want to keep trying."[42]

Practically from the start, the Cadys noticed that Ruthie and Verena were developing distinct personalities. By age five, Verena was more verbal and more cautious, while Ruthie had become more domineering and mischievous and tended to like hands-on activities. Verena loved eating, while Ruthie found it boring, so that Verena tended to eat for both of them for as long as Ruthie would sit still.[43] What about disciplining attached children? "When one misbehaves," Marlene admitted, "of course the other has to suffer some of the consequences too. Ruthie has had to go stand in the corner a few times for being naughty, and Verena just goes along with her. She'll say, 'You can't leave the corner yet, Ruthie, you're still naughty,' even though she is there too. She just understands that Ruthie has got to be punished and that's the way it is."[44] On the other hand, "They're never lonely, they always have someone to hug them if they get hurt or to share secrets with."[45]

Two years before the girls' deaths, Marlene recorded her impression of the twins' attitude toward surgical separation: "If someone asks if we have any plans for surgery, the twins find that very upsetting." Marlene and Peter had already decided that if separation were ever offered, Ruthie and Verena would have to decide for themselves whether to pursue it. "We have to remember that it's not just a matter of separating them physically," Marlene declared, "but also psychologically, spiritually, and emotionally. I don't know that the rest of us really have the capability to make that decision for them."[46]

As it turned out, the twins grew more and more ill as a result of their anomalies. They died, still conjoined, at the family's cabin with their parents and sister around them. Ruthie became quiet just before her death, while Verena comforted her and gave Marlene a list of friends to whom she wanted flowers sent in their names. She also named the people who were to be invited to the funeral. "Ruthie and Verena taught us so much," their mother concluded. "They were a perfect example of sisterhood and unity."[47]

In 1949, Yvonne and Yvette McCarther were born to Willa Jones in Los Angeles. Joined at the head, the girls were kept in the hospital and studied for two years as doctors tried to figure out how to separate them. The girls' brains were distinct, "but their circulatory systems were linked at their heads in a way that made surgery impossible."[48] Willa, "fearing surgery would be fatal to one or both, refused to allow an operation."[49] When her daughters reached the age of two she was advised to institutionalize them, since doctors predicted that they would never walk and would need life-long nursing care.[50]

Willa chose instead to bring her children home and to teach them "to think of themselves as individuals."[51] For most of their lives, Yvonne and Yvette enjoyed a low-profile existence, although when they were small children their mother reluctantly decided to exhibit them with the Clyde Beatty Circus in order to pay their enormous hospital bills. (In the case of conjoinment, as with other unusual anatomical conditions, vast sums of money and other resources seem to be available for "normalization" but very little for ongoing support, particularly for twins who remain "uncorrected.")[52] Yvonne and Yvette not only learned to walk, thereby defying medical predictions; they also earned high school equivalency diplomas in 1967.[53] They were finishing their associate nursing degrees in 1993 when they died one night "of natural causes," in their own apartment, at the age of forty-three.[54] The twins had been pursuing nursing degrees because they believed they could do good work with handicapped children.[55] Their funeral was well attended by their friends and acquaintances, who had appreciated them for their lively sense of humor, open-mindedness, and profound "zest for life."[56]

For a brief period in the 1970s, Yvonne and Yvette McCarther traveled around the United States performing with fellow gospel singers, and in doing so they mirrored, presumably unwittingly, the lives of two other African-American conjoined sisters born a century earlier. Millie and Christina McCoy, billed as the "Two-Headed Nightingale," were born to enslaved parents in North Carolina on July 11, 1851, joined "from the lower ribs to the bottom of their trunks."[57] Despite a harrowing childhood in which they were repeatedly sold, traded, kidnapped, and displayed, the sisters seem to have grown into reasonably happy, healthy, and well-educated adults, thanks in part to their final, relatively decent caregiver, Mrs. Joseph Smith.[58] Even after legal emancipation, the formerly enslaved twins opted to stay with Mrs. Smith, apparently out of genuine affection and appreciation, and they went on to earn large sums of money via their performances, which included singing and dancing.[59] Indeed, legend has it that the sisters earned enough to enable their father to purchase "the very plantation on which he was once a bondman, and on which Millie [and] Christine first saw the light of day."[60] Numerous first-hand accounts described the sisters as intelligent, pious, and cheerful—and as very good singers and dancers.

Abigail and Brittany Hensel, Americans of European descent, are widely known through portrayals in *Life* magazine and family interviews on TV. Born in 1990, they are twins of the dicephalic type: their body is largely like that of a singleton, with two arms and two legs (a vestigial third arm was removed when they were babies) but with two heads. Doctors believe Abigail and Brittany have two sets of most internal upper-body organs (including two hearts) and one set of most internal lower-body organs. The girls live a fairly quiet existence with their family in the Midwest. Aside from one relatively brief hospitalization for pneumonia, "the girls have required no extraordinary medical care."[61] "The idea of separating the twins was dismissed by both parents right from the start, when doctors said there was little chance that both could survive the procedure."[62] They and their parents, Patty and Mike Hensel, agree that separation surgery would present unnecessary risk. "If they both lived [through the separation]," Mike asks rhetorically, "what kind of life would they have? They'd be in surgery for years, suffering all the time, and then they'd have half a body each."[63]

In choosing to resist medicalizing their daughters' unusual anatomies, Patty and Mike show that they are rather extraordinary—in some ways more extraordinary than their (by all first-hand accounts) rather typical children. Patty's experience as an emergency-room nurse may well inform her understanding of the dangers involved in seeking simple medical "cures" for complex problems. But whatever the reason, Patty and Mike seem to view their daughters precisely the way they do their other children—as individuals deserving the same support, discipline, and love. They encourage them all to express their individuality in terms of hobbies, clothing, tastes, and friends. As a consequence, the twins are doing very well. They walk, play, go to school, swim, bicycle, draw, learn, and have dreams for the future. As one might expect, occasionally someone stares or exclaims astonishment at the sight of them, but they take this in stride. Even when they were very young, they had individual aspirations. By the age of six, Brittany had declared her hope of growing up to be a pilot and Abigail was saying she wanted to be a dentist. A reporter for *Life* who spent time with the Hensel family noted—like nearly everyone who has commented about them—that "an unmistakable air of well-being suffuses this household."[64] (See Figure 5.)

Many singletons, on meeting conjoined twins for the first time, express amazement at just how *normal* and *healthy* they seem in every way but the obvious. Indeed, Chang and Eng's publicity material often sought to surprise the viewer with their normality (see Figure 6). Yvonne and Yvette McCarther's college "classmates [were] said to treat them no differently from anyone else."[65] For their own part, the McCarther twins described themselves as happy; they did not "regard themselves as handicapped or deformed but merely different."[66] Similar things were said about the conjoined Burmese boys Lin and Win Htut, who in 1984, at age two and a half, were admitted to a Canadian hospital for separation surgery. "Once we overcame our initial reaction to their deformity," the nurses assigned to the twins in the intensive-care unit remarked, "we were struck by their normalcy." In fact, before the operation, "as nurses we were not sure what to do with [these] 'healthy' children."[67] But the nurses were deeply troubled after the operation: "The healthy 'whole' children whom we had adopted as our own were now, seventeen hours later, separate but badly deformed. *Now* they seemed handicapped."[68]

How do conjoined twins cope with their attachments? Like the rest of us who live in commitment with others, they work out explicit and tacit agreements about day-to-day living. For instance, Chang and Eng "had

5. Abigail and Brittany Hensel playing at home with their mother and brother, in a photo from 1996.

"CHANG" AND "ENG"
THE WORLD RENOWNED UNITED
SIAMESE TWINS.

6. Eng and Chang Bunker engaged in their various pursuits—farming, fishing, boating, spending time with family, and so on—as shown in a mid-nineteenth-century lithograph by Currier and Ives.

agreed that each should in turn control the action of the other. Thus Eng would for [a set number of days] be complete master; they would live for that time at Eng's house, and Chang would submit his will and desires completely to those of Eng, and vice versa."[69] Similarly, by the age of seven Ruthie and Verena Cady had "solved the who's-in-charge problem" by deciding that they would take turns making the important decisions, each on alternate days.[70] Abigail and Brittany Hensel walk, run, and swim in harmony by a system not fully understood by doctors; each eats the food she prefers, and they (often unconsciously) negotiate activities which require the participation of both. Indeed, many conjoined twins seem to consider themselves *better* prepared than singletons for the rigors of the world. Chang and Eng, for instance, were said to have "learned to accommodate themselves to their situation; and probably they regarded themselves as equally, if not more favorably situated in respect to the necessities of life than if they had enjoyed a separate existence. They brought to the accomplishment of any undertaking, if needed, a double strength and a double will."[71] Similarly, conjoined sisters Mary and Margaret Gibb (1912–1967) "came to prefer their condition over the alternative of separated lives."[72]

So conjoinment does not automatically negate individual development and expression, any more than other forms of profound human relations do. Indeed, differing personalities and tastes are the rule among conjoined twins with two conscious heads.[73] As J. David Smith noted in his psychological review of the histories of conjoined twins, the fact that such twins invariably have distinct personalities confounds simplistic responses to the nature/nurture controversy; people who are conjoined appear to share the same nature (genetics) and nurture (environment), yet end up as unique individuals.[74] Chang and Eng were always said to be "perfectly dual in mind," though they both loved chess, reading, and riflery.[75] Late in life, Chang reportedly engaged in "immoderate drinking," a habit Eng apparently did not share.[76] The Czechoslovakian sisters Rosa and Josepha Blazek (1878–1922), known as the Grown-Together Twins, "in the manner of food and drink . . . had different likes and dislikes, as well as . . . different impressions of people and other subjects."[77] Lori and Reba Schappell are so obviously different in personality and tastes that when I interviewed Lori I found myself wondering aloud whether she and Reba would be friends if they were not conjoined. Lori thought about it for a moment and answered that she thought they would, but they probably wouldn't go shopping together. (Reba likes to shop efficiently according to a prepared list, while Lori is a browser and an impulse buyer.)[78] Abigail and Brittany Hensel also have markedly different tastes and personalities. And the list goes on and on. Most conjoined twins appear to favor speaking in the first person singular—that is, each speaks for him- or herself. It seems to be typical for each to think of him- or herself as a unique, individual being.

In this sense, Ladan and Laleh Bijani of Iran were like most conjoined twins. Having lived conjoined for twenty-nine years, they had developed psychosocially as two distinct individuals, with notably different personalities, interests, and tastes. Attached to her sister near the top of her head, Ladan, the more outgoing and assertive of the two, found fulfillment in the study of the law. Meanwhile, Laleh, a more demure personality, took pleasure in playing video games and caring for animals.[79] In their differences, the Bijanis sounded like the Bunkers, the Blazeks, and the McCarthers. But the Bijanis were unlike all their predecessors in one important respect: they decided that their conjoinment intolerably limited their lives. And so in 2003 they were the first twins in history to be separated by an operation to which they had personally consented.

Indeed, Ladan and Laleh more then consented: they aggressively sought out the surgery, traveling the world to find surgeons willing to undertake it. In 1996 a German team examined them closely, at their request, and determined they were too intimately joined to be separated safely. But Ladan and Laleh pressed on. When in 2001 they heard that Ganga and Jamuna Shrestha, a pair of Nepalese conjoined children— also joined at the head—had survived separation at the hands of Dr. Keith Goh of Raffles Hospital, a private institution in Singapore, they went to Dr. Goh and asked him to separate them.[80] Ladan explained that she wanted to pursue law in their hometown of Shiraz; and Laleh —though she had taken a degree in law alongside her sister—wanted to become a journalist in Tehran.[81] Ladan insisted, "If our situation goes on [like this] for one or two more years, we wouldn't be able to stand it. We are two completely separate individuals stuck to one another, with different world views and lifestyles."[82] Throughout the drama, the international press reported the women's poetic claim that they just wanted to see each others' faces without a mirror[83] (the way they were conjoined, near the back of their heads, made this impossible), but it was clear their goal was really to live physically distinct.

Dr. Goh, like previous observers, was struck by how seemingly happy, educated, and well adjusted the Bijani twins were. Because of this, and because the surgery was clearly high-risk and optional, he and his team were reluctant to pursue the operation. Dr. Goh told reporters he repeatedly tried to talk the women out of the operation.[84] Nevertheless, the path was cleared by an internal ethics committee made up of physicians and the hospital's mental health professionals, who decided the women truly knew what they wanted. For his part, Dr. Goh grew convinced that, appearances to the contrary, these seemingly happy women "suffered a deep but concealed misery."[85] He felt he had to do what he could "to give these girls some measure of a decent, normal life as we know it."[86] So plans went forward. Just before the surgery, the twins wrote a letter of thanks to their well-wishers across the globe, a letter that Raffles Hospital posted on its website. God willing, the sisters told the world, they would soon be living separate lives. But the story ended in tragedy: fifty hours into the surgery, just after the final cut to separate them, both sisters hemorrhaged uncontrollably. Ladan died first; Laleh an hour and a half later. Their bodies were sent home to Iran in separate coffins. Saddened and shaken, Dr. Goh told the press, "At least we helped them achieve their dream of separation."[87] Dr. Loo Coon Young, chairman of the Raffles Hospital, concluded, "As doctors there is only so much we can do. . . . The rest we have to leave to the Almighty."[88]

At least one man angrily refused to blame anyone but the doctors for the sisters' deaths. Laleh and Ladan's adoptive father, Alireza Safaian —himself a physician—had been convinced all along that separation would mean certain death. Never believing the separation surgery was in his daughters' best interest, Dr. Safaian told the press, "They were victims of a big propaganda [campaign] in Iran and Singapore." Now beside himself at the loss of his daughters, Dr. Safaian insisted "the twins had led normal lives before the surgery." Far from being entrapped by conjoinment, they had had their own apartment, "did their own shopping, and cooked for as many as twenty guests." They had had friends, education, and the basic comforts of life. In their father's eyes, Ladan and Laleh had been two healthy, well-off individuals, killed without reason.[89]

But most public commentaries on the Bijanis' deaths differed markedly from the views of Dr. Safaian. Even while mourning Laleh and Ladan's loss, commentators lauded their heroism and bravery and applauded the doctors' willingness to follow their patients' wishes.[90] Few singletons could imagine living a life conjoined, and most believed the sisters had made the right choice—risking their lives to try to achieve physical independence.[91] But few recognized that this choice—to risk their lives to achieve an embodiment radically different from the one they had been born with—made Ladan and Laleh quite unusual, unusual even for conjoined twins. This is because most such twins, like most of us singletons, grow up accepting the basic bodies they were born with as necessary to their selves. Most people who are conjoined, given the opportunity to do so, accept and embrace a life of two minds in one packaging of skin.

Indeed, many conjoined twins often explicitly say they do not want ever to be separated, since this would result in a profound change of identity or the death of a twin's "other half."[92] Imagine having a vibrant, cherished, and articulate part of your body amputated and lost forever, or cut loose to lead an independent life! Even as late as 1869, Chang and Eng had "no desire to be surgically divided from each other . . . [although] some of their relatives have become anxious that they should be separated, if it were possible to do so; for latterly their two families have been living apart."[93] Abigail and Brittany Hensel, Reba and Lori Schappell, and others have repeatedly said they would never wish to be apart. One might assume that these sorts of claims are simply psychological coping mechanisms—that conjoined twins "decide" they do not wish to be separated because this way the necessary attachment seems like a choice. But then is it a mere coping mechanism when a woman says she's comfortable being female, or when a Native American says he's comfortable being Native? Historically, the medical profession has viewed as psychologically *ill* those individuals—such as transsexuals and people seeking the amputation of healthy limbs—who have longed for bodies very different from the ones they were born with.[94] Why not see conjoined twins who wish to remain together as well adjusted?

Experts on conjoined twinning have sometimes turned to the psychological literature regarding twins born separate and have argued that "the unique difficulties [all] twins have in creating ego boundaries and a sense of individuation" are a major reason for attempting separation of conjoined twins at all costs.[95] But there are two problems with this line of reasoning. First, the evidence tells us that conjoined twins who have remained conjoined do in fact become individuals in the psychological sense, if not in the physical sense; each speaks of him- or herself as an individual, and they develop personalities and tastes distinct from those of their siblings. Second, the notion that a conjoined twin must individuate to the same degree singletons do takes singleton development, unjustifiably, as the standard for everyone.[96] Indeed, after reading many biographies and autobiographies of people who are conjoined, one has to wonder whether we might not *all* benefit from more twin-type behavior in this world—that is, whether we might not all benefit from a little *less* "individuation." Many conjoined twins are models of cooperative behavior, thanks to their lack of total individuation. Abigail and Brittany Hensel are good examples of this: "Teamwork is a concept [they] have grasped more quickly than their peers. Once, after several students got into an argument, the twins led a class discussion on how to get along."[97]

Remarkably, I have found only one published report of a face-to-face psychological evaluation of conjoined twins which claims they suffered profound distress from their conjoinment. The article, which appeared in the *Journal of Projective Techniques and Personality Assessment* in

1964, details the results of a series of psychological tests performed on twelve-year-old twin boys. "Practically their entire fantasy life," the psychologists stated, "is consumed by the completely frustrating situation in which they find themselves, locked together for life."[98] Some of the evidence: when shown Rorschach inkblots, the boys saw "Siamese twins joined together," and when asked to draw a person, "each [drew] two figures united by a common bond."[99] Of course, these results might have indicated *not* frustration with the conjoinment but merely an awareness of it, or even a tacit acceptance of it. Wouldn't we be surprised if a singleton child, when asked to draw a person, drew conjoined twins? It is true that the boys also seem to have revealed dark, depressed feelings, and perhaps some physical hostility to the anatomical bond itself. One of the boys "pounded furiously on top of the [flesh] bridge exclaiming loudly, 'See, we are joined together forever, we were born that way and we will always be that way.'" There seemed to be a "profusion of hostile content in both their inkblot fantasies and TAT stories, where people [were] described as mad and murderous and objects as exploding and burning."[100] Yet the psychologists readily admitted that the boys had been treated very poorly throughout their short lives: "The twins cannot be said to have lived anything approaching a normal family life. Nearly the entire first two years of life were spent in a hospital where special care and observation were necessary. The next two years the twins did live in the family home, but according to the father . . . the mother's reaction to this grotesque and irremediable abnormality was violently rejecting. . . . [The father took] the twins out of the home and [put them] on exhibit beginning at age four."[101] The boys subsequently lived in a small trailer and apparently never received any formal education. At age twelve they were illiterate, and had no prospects of learning to read and write. When they weren't being exhibited at the circus or examined by medical personnel, they were left in the trailer, where they watched television almost constantly.[102] Is it surprising that they would feel angry about their situation?

Notably, the authors of this 1964 report closed their discussion by admitting: "A search of the medical literature reveals that conjoined twins who survived to adulthood are reported to have refused surgical separation in every case, even when one of the twins was dying and separation would have permitted survival of the remaining twin."[103] Yet the authors implied that such an attitude must represent a pathology or a failure of reality testing, rather than an understandable conscious choice or a normal psychological stance.

The available documentation shows that the desire to remain together is so widespread among communicating conjoined twins as to be practically universal. In other words, people who are conjoined and able to communicate seem to be almost as disinclined to be surgically separated as singletons are to be surgically joined. The Bijanis' remains the only case in which conjoined twins old enough to express preferences have consented to a separation. Moreover, conscious conjoined people whose twins have died have invariably chosen to remain attached, knowing that this means they will soon also die, and that in the interim they will be attached to their dead sibling. The sisters known as the Biddenden Maids, born in England in 1100, lived to the age of thirty-four. When one died, the survivor was offered separation surgery but refused, saying, "As we came together we will go together."[104] Records indicate that the survivor died six hours later. Millie and Christina Mc-Coy, "having lived thus long together, . . . express[ed] no desire to be parted and hope[d] to leave this world as they came into it—together."[105] Eng does not appear to have explicitly asked for separation when Chang died, though he "had continued rational" after becoming aware of Chang's death. When told by his son that Chang was dead, he reportedly declared, "Then I am going!"[106] Eng seems to have acknowledged that as they were joined in life, so they would be joined in death; and this was evidently the sentiment of the surviving twin in the case of Millie and Christina, Ruthie and Verena Cady, the Biddenden Maids, and all the others whose histories are recorded. When, in 1967, Margaret Gibb was diagnosed with cancer, she and her sister Mary "dismissed completely the idea that they should consider surgical separation, and the cancer rapidly spread to Mary. They died within two minutes of each other."[107]

It is safe to conclude, then, that conjoinment becomes so essential to these twins—to their sense of who they are—that they cannot readily conceive of living in a different mode. And this fact resonates with the claims of many other people with stigmatized identities. When I asked Ruta Sharangpani if she could imagine herself not having a congenital visual impairment, she answered that, though it would be easier in a practical sense, it was unclear who she would be without it: "It's something that runs so deeply for me, just as my race, my Indian-ness is something I can't imagine not being. I can't think of not being a woman. I am Indian. I am visually impaired."[108] The actor Camryn Manheim makes a similar observation in her autobiography, *Wake Up, I'm Fat!*, concerning the way in which body size becomes an aspect of one's identity: "When I'd lost the weight, I had lost myself."[109] As repulsive, sad, pitiful, and *unnecessary* as these conditions may seem to someone who does not inhabit them, they often function as an inexplicable, essential, even cherished aspect of the self for those who do inhabit them. And they often come to be an inexplicable, even cherished aspect of particular people in the eyes of those who know them. So, for example, a number of parents of children with cleft lip have confessed that, when their children's cleft lips were repaired (typically several months after birth), they missed the cleft and quietly mourned its loss. Joanne Green, mother of three children born with clefts and the founder of a parental support group, warns other parents about this: "Very few parents are initially thrilled with the surgery. The baby will almost seem to be another baby. There will be a marked difference in the face. And it will take you a while to adjust to this new face. After all, you loved the old one!"[110] Even knowing in advance that the cleft lip will go, parents often come to see it as an essential part of their children, the same way they view their children's sex. These things are, after all, *con-genital*— that is, happening with (con) the very genesis of the child and the child's identity. This con-genesis is probably one reason people find it easier to accept their anatomical differences when these are present at birth, compared to stigmatizing variations resulting from postnatal illnesses and accidents.[111]

Of course, most people with unusual anatomies do not view their differences as something that shapes their every moment and colors their every thought. "I'm a conjoined twin because it happened at birth," says Lori Schappell, "but I do not live a conjoined life. I don't think of it every minute of the day. The only time I think of it is when I'm interviewed. It's just an integral part of my life."[112] What we learn, then, from people who grew up conjoined is not that it is a state absolutely preferable to being a singleton; rather, it is a state that the conjoined have generally been able to accept and view as a seamless part of their identities.

Read closely, the biographies and autobiographies of conjoined twins reveal that the limits on these people's freedom stem chiefly not from their conjoinment per se but from the thoughts and actions of others. The story of Daisy and Violet Hilton seems a good example of this. Born in 1909 in Brighton, England—either to an unwed mother or to a married couple who died soon after the birth—they appear to have spent most of their youth as the property of greedy managers. They were connected by a bond about fifteen centimeters in diameter at the base of their spines, and doctors repeatedly said they would die if separated surgically. Though their movements were at first constrained, the connection became more flexible over time and they were able to move about quite well by the age of two.[113] Groomed to be performers, they worked on the entertainment circuit from an early age, without much control over their lives or the money they earned. In 1932, after successfully suing their managers, they were finally awarded independence and "one hundred thousand dollars in damages."[114] For a time, Daisy and Violet continued their performances, appearing in vaudeville reviews and two Hollywood films (*Freaks* in 1932; *Chained for Life* in 1951), learning the dance known as the Black Bottom from Bob Hope, and traveling widely throughout the United States.[115] In 1934 Violet's attempts to marry her musical director, Maurice Lambert, attracted national attention as the couple wandered through twenty-one states trying to get a marriage license. This was repeatedly denied them, on the grounds that to give a conjoined twin a marriage license would be immoral. Eventually both Violet and Daisy did marry, in 1935 and 1941 respectively, but neither marriage lasted long, for reasons that are unclear. Over time, either Violet and Daisy grew tired of the show business life, or the public stopped being interested in them, and by 1960 they were working in Charlotte, North Carolina, as supermarket clerks. They died in 1969, of complications arising from the flu.[116]

The curiosity and condemnation people expressed about the Hiltons' sex lives seems to have been more strident than usual; but such reactions have always been associated with conjoinment. Many singletons simply cannot abide the idea of conjoined twins having sex. So what *do* conjoined twins do about sex? So far as I can ascertain, they do what most people do: seek out a lover, find a little privacy, and do the deed, understanding it to be a one-on-one affair. Of course, the other twin is right there, but people who are conjoined who have talked about their sex lives consistently report that during sexual intimacy their siblings remain quiet and mentally distant. Supposedly, Harry Houdini taught his friends Daisy and Violet "how one could withdraw mentally while the other engaged in the pleasures of sex."[117] The Bunkers implied that they used a similar system; Lori Schappell, likewise.

Obviously such sexual relations are unusual, and they no doubt are partly to blame for the fact that conjoined twins seem to have a hard time establishing long-term romantic relationships with others. But another possibility is that people who are conjoined obtain from their twins something akin to the stalwart companionship, understanding, and unconditional love many others find primarily through committed romantic partnership. *Side Show*, a musical based on the lives of Daisy and Violet Hilton, captures this possibility beautifully in the closing duet, "I Will Never Leave You." Though the characters of Daisy and Violet sing it to each other, it is a romantic number that would sound just as appropriate being sung by a couple of lovers. Emily Skinner, who played Daisy in the Broadway production of *Side Show*, understood this: "I watched this show on CBS called 'Twin Stories' recently. There were these Siamese twins talking about having grown up as twins, and they said it's like being born with your soul mate. I thought, 'That's so true.'

And that moment at the end [of *Side Show*]—when they sing 'I Will Never Leave You' and have that spiritual moment where they accept themselves and become strong because of it—that moment is like saying, 'You are my soul mate, and I love you.'"[118]

When I was working in 1999 with the director Ellen Weissbrod on a documentary about Lori and Reba Schappell,[119] Ellen pointed out to me that when you immerse yourself in the thought of conjoinment for a long time, you begin to hear every crazy-in-love song as a song about conjoinment:

> I've got you under my skin
> I've got you deep in the heart of me
> So deep in my heart
> You're really a part of me
> I've got you under my skin . . .
> I would sacrifice anything
> Come what might
> For the sake of holding you near.[120]

Songs about never being alone, songs about feeling the constant touch of another, songs about someone who knows you as well as you know yourself—all of them sound like celebrations of conjoinment. A singleton is apt to find this discovery very disconcerting. No matter how much they resonate, these age-old effusions about attachment are intended to be just metaphorical. Love is supposed to be an experience of the individual, and an individual is expected to be the only inhabitant of his or her skin. But people who are conjoined stretch the limits of individuality. They lead the thoughtful, sympathetic singleton to consider the degree to which any of us truly are or wish to be independent of others, and to ask why individuality—or any other aspect of humanity—need be thought of as limited to one particular kind of anatomy.

Split Decisions

2

The earliest known attempt to separate human conjoined twins occurred in Byzantium roughly a thousand years ago. According to a chronicler of the tenth century, it involved twin boys "connected together from the Ombilic down to the lower abdomen, in a position face to face."[1] As with nearly all surgical separations before the twentieth century, this one was attempted because one of the twins had died. The bid to free the survivor failed, however—he perished three days after the operation.[2] The first nonlethal separation appears to have been the one performed in 1690 by the German anatomist König, who divided infant girls joined by a small abdominal band of flesh. König is said to have "applied a ligature to the middle of the connecting band, tightened it daily, and, at last, successfully divided the remnant of the band with the knife."[3] By 1964, at least twenty-six surgical separations had been attempted and the pace of nonemergency separations was increasing.[4] The count for surgical-separation attempts is presently up to about 250, though the actual figure may be higher, since, as one surgeon has noted, "potential authors have an understandable reluctance to report their failures."[5]

Given their rarity and difficulty, separation surgeries leave medical participants with intense memories and emotions. Pediatric surgeon Mark Stringer, Fellow of the Royal College of Surgeons, vividly recalls the separation of three-year-old twins Katie and Eilish Holton. It was performed in 1992, near the end of Dr. Stringer's formal training.

I was a senior registrar on the team. I was very conscious that I'd had no prior experience in this area. I was anxious about how it was all going to go. I was anxious about whether we were doing the right thing for the family—that is, whether we could achieve what the parents were hoping we could. I was very much the person on the floor coordinating the investigations being requested [in preparation for the surgery].

I developed quite a relationship with the girls, which was also quite difficult. I got very fond of them. They were both very much individuals. Three is a very nice age, from my point of view—there's a lot that goes on. They were very intelligent girls; we had a good level of communication; we played a lot of games. They were going through a miserable time in hospital [prior to the operation]. You need to bond with the kids when you're their doctor, so they can trust you. I got quite close to them and to the parents.

As time goes by—seeing events unfold, having time to reflect—I feel a little uneasy about whether we did the right thing, but I know the parents only ever had the best interest of the girls at heart. . . . I had doubts before [the separation], and I really had more doubts afterwards. That is in no way to undermine the people I was working with. I think Lewis Spitz [head of the medical team] is a superb surgeon. I felt we were going about things in a very structured, very careful manner. . . . The enormity of what we had done [only] hit me afterwards.[6]

Katie and Eilish had been born in Ireland in 1988, and were pretty healthy, all things considered. They had spent the first four months of their lives in the hospital so that doctors could try to figure out their anatomies and consider whether separation would be possible. But afterward their parents, Mary and Liam Holton, took them home to join their singleton siblings, regarding Katie and Eilish "as normal children in a unique situation."[7] The girls were given regular physiotherapy treatments to help expand their range of movement. A documentary film that includes pre-operation footage shows "their loving acceptance and participation in the 'everyday' life of their family and community."[8] Indeed, one doctor who observed the family's obvious psychosocial well-being remarked: "These twins are going to have great difficulty believing that there is really anything very much wrong with them."[9]

But there was something very much wrong with Katie and Eilish by conventional standards. The upper part of their body looked like two girls, while the lower part looked like one. It was clear that separating the twins would constitute a major undertaking. According to the medical report published in 1994, "they were joined from the forearms and shoulder to the pelvis . . . and faced each other obliquely at approximately 120 degrees. Their conjoinment circumference measured 78 cm. They had two independent, normal lower limbs, two normal upper limbs, and independently functioning but conjoined upper limbs."[10] While Katie and Eilish had a total of two hearts and two sets of lungs, they shared much of their lower gastrointestinal tract and reproductive system. As often happens with children born conjoined, they also showed signs of asymmetry. In this case, Katie's spine was curved with scoliosis, and her head tended sharply to the side. Their inability to walk without aid seems to have been a major factor in Liam and Mary's decision to pursue separation. Another, undoubtedly, was the desire to give the girls normal-looking bodies.

According to the physicians, Katie and Eilish "were healthy and intelligent" when admitted to the hospital prior to the surgery.[11] It is obvious why their parents and doctors had weighed the separation question very carefully. The surgery was unlikely to improve their fairly good health; in fact, it would leave them with significant impairments. Each would end up with only one leg. Each would have to learn to walk with a prosthetic leg, and then increasingly larger prosthetics to keep up with growth over the years. Each would be given half a reproductive system, including half a uterus, half a vagina, and half of their innate external genitalia. This would reduce if not eliminate their fertility and probably diminish their sexual sensation. One twin would get the bladder and urethra, the other the anus and rectum, meaning that each would have to wear a collection bag for urine or feces. The conjoined arms would be divided between the girls, but they would not look or function the way arms usually do. The chest-to-pelvis wound left by the surgery would be enormous.[12] And these constituted the *minimal, guaranteed* negative effects of separation. Plenty of other things could go wrong.

One of the greatest challenges in the separation of conjoined twins consists in closing the surgical wounds and staving off infection. To prepare for this challenge, surgeons admitted Katie and Eilish to the hospital months before the separation and inserted tissue expanders under their skin. These promoted the growth of new skin, which could be used after the separation to close the wounds. "This was well tolerated," according to the physicians' report, "but supplementary nastrogastic tube feeding was required because of anorexia and weight loss."[13] In other words, the skin expanders were not rejected by their immune system, but the girls lost their appetites and started losing weight. The medical staff had to begin feeding them through tubes inserted in their noses and down their throats. Dr. Stringer explains that the loss of appetite was partly physical in nature: "As you expand the abdominal cavity, there is a physical restriction to appetite from a space-occupying effect."[14] That is, abdominal expanders can make the patient feel full. But there was probably also a psychological reason for the loss of appetite:

> These twins were outside their normal environment, stuck in hospital, a lot of people coming and going. That must have been very distressing for them. . . .
>
> There were a lot of attempts to talk to them, to put things into their language, [to tell them] what was happening. We weren't going about it in an authoritarian way. We tried to engage them at all sorts of levels: play therapists, psychologists, and with nurses who had been through it before. We spent a lot of time with the parents and with Katie and Eilish themselves, trying to help them understand what we were doing.
>
> [But] there must have been a psychological aspect to their anorexia, as well as physical. They weren't active [they were largely immobilized for the preparatory procedures], and they were being starved [subjected to prescribed fasting] for various procedures—for anesthetics, scans, etc. All that takes a toll.[15]

The surgery took place on April 1, 1992, by which time the girls were three years and seven months old. The operation lasted fifteen hours.

Katie died four days later. Doctors ascribed her death to myocardial insufficiency—essentially, heart failure. According to Stringer, the girls' hearts were "enormously stressed" by the surgery,[16] and apparently Katie's had been relatively weak. (The doctors did not know this prior to separation.) So—one might assume—without separation Katie, and perhaps also Eilish, would have soon died from cardiac insufficiency. But such an assumption isn't at all warranted, since Katie and Eilish's hearts working together might have kept them both alive for a very long time. Despite numerous tests and examinations, the doctors had observed no signs of trouble with Katie's heart until the surgery.[17]

After Katie died, surgeons harvested some of her skin and used it on Eilish in an effort to improve Eilish's health and reduce her chances of infection.[18] Eilish's psychological distress at the loss of her sister was evident early on. She named her prosthetic leg "Katie."[19] Today she is able to move about quite well and no longer looks anything like a conjoined twin. She is disabled only in the more conventional senses of the term.

We normalize our children in so many ways. I'm teaching my son when and how to say "please," "thank you," and "you're welcome." I'm teaching him standard American English, how to use verb tenses correctly, who should be referred to with the pronoun "he" and who with "she." I'm trying to get him to eat balanced foods so that he'll grow in size and strength. I try to get him to sit up at the table and look people in the eye when speaking with them. I've even chosen medical forms of anatomical enhancement for him—namely, vaccinations against diseases.

When a child is born with an unusual and potentially stigmatized anatomy, the parents' desire to normalize that child can be especially strong. Most children with unusual anatomies are born to parents who do not share the unusual trait, and so the parents' reaction often involves fear, confusion, shame, guilt, and distress, even while those feelings are tempered by relief, excitement, and joy at the birth. The parents often can't imagine living "that" way. They flash back to the worst teasing they ever experienced when they were young, and imagine it being infinitely worse for this child. They remember how difficult it was to make friends and how much personal appearance counted, and they worry that this child will always be alone. They know how important it is not to feel alone.

Parents of newborns with unusual anatomies often have a hard time seeing a socially significant anatomical difference as anything other than a medical problem, and indeed sometimes these conditions do come with serious physiological concerns. For example, cleft palate, a condition in which the roof of the mouth doesn't close fully during prenatal development, often complicates feeding and increases the risk of significant ear infections. One form of intersex, Congenital Adrenal Hyperplasia (CAH), entails an underlying metabolic imbalance that can be serious. But parents also tend to see anatomical anomalies as medical problems because U.S. culture tends to see nearly everything anatomical as a medical issue. Weight, hyperactivity, circumcision, menopause, depression, birth, death—all of these things, for better or for worse, have come to be seen primarily as individual medical issues. In this context a child with an unusual anatomy seems to be a child who, regardless of her actual health, is first and foremost in need of a doctor to "cure" her.[20]

Parents (and the general public) often imagine that surgical treatments of congenital anomalies will be a "quick fix"; they believe this despite warnings from experts that such surgeries often carry serious costs and risks, may never make the child look "normal," and sometimes require repeated follow-up procedures. Many parents want a surgical fix not only because they wish to prevent lengthy suffering (even though the child may not actually be suffering at all), but also because (as noted in the case of dwarfism) parents who were expecting a typical newborn often go through a grieving process.[21] They grieve for the "normal" child they expected and feel the loss quite acutely. In seeking a surgical appearance-normalization, they are hoping for what Arthur Frank has called the "restitution narrative" in medicine: that the perfect child they imagined was theirs will be "restored" to them through medicine, specifically surgery.[22] As parents come to know, love, and bond with the child, sometimes this desire for "reconstructive" surgery fades. But not always. Indeed, the more parents come to believe through familial bonding that the child—despite appearances to the contrary—really *is essentially* perfect and normal, the more they may want surgery, because it seems to promise to make that perfection and normality obvious to any clod who might think otherwise. Far from feeling like a rejection of the child, normalization surgery may feel to some parents like a manifestation of full acceptance and unconditional love.

But parents may also seek surgical fixes because of the genuine (and typically unaddressed) anguish they are feeling about their own identities. For when a child is born with an unusual anatomy, not only is that child's identity thrown into question, but so are the identities of his parents.[23] Parents whose child displays an uncertain identity—whose child may not be clearly one or two, or may have an unfamiliar sort of face, or may have genitals that don't look like the usual male or female—these parents suddenly find themselves unsure about their *own* social and familial role. How are they supposed to act? What are they supposed to think, feel, do, say? They know only how normal parents are supposed to behave, but they can't be normal parents if they don't have a normal child. They seek surgical "reconstruction" of a normal child in part because they feel like they will know how to be a parent to that child, whereas they often feel uncertain how to be a parent to this one.

Now, doctors are supposed to fix uncertainty by providing certainty; certainty is much of what patients seek from them. Technically, we go to doctors for drugs and procedures, for interventions. But I think most of us actually go for diagnoses and prognoses—for certainties. Uncertainty is perhaps the scariest thing in a medical setting; it is treated as even a bigger enemy than death. And Americans typically see medical interventions as the only appropriate way to deal with anatomical uncertainty: more tests, more procedures, even if that is riskier than refraining from tests and procedures. So when a child is born with an unusual anatomy, doctors often rush to offer surgical normalizations, reassuring the parents that the child can be made more normal. They understand this to be their role, as doctors, as pediatricians, as compassionate human beings. Some believe they must offer appearance-normalizations to enable the parents to accept and bond with the child.

Today, surgeons can often separate conjoined twins without loss of life to either child. They can often rebuild cleft palates, improving function, and cleft lips, improving appearance. They can often lengthen the limbs of children with achondroplasia, so that these children end up notably taller, more like other people in height. They can often construct less confusing-looking genitals for children born with intersex conditions, so that big clitorises end up looking more like petite clitorises, with labia and "neovaginas" where there were none before.

So why shouldn't we give every child a shot at life with a normal-looking anatomy? Some critics might argue that these kinds of normalizations reduce visible diversity, and that visible anatomical diversity is inherently good—perhaps it has the potential to breed understanding and compassion, to provide variety and richness. Normalizations would therefore be inherently bad. Yet this sort of claim, when made in reference to a real child, often feels to me like a sacrificing of that child on the altar of social progress. I suppose it might be the case that anatomical diversity is inherently good—if enough of it were visible, it might force us all to recognize the difference between hierarchy and variation and the reasons for each—but I can't say I would always choose to have my own kid fight that awfully big battle. We'd do well to acknowledge how all of us choose minor anatomical normalizations each day, either for ourselves or for our children, and to ask why we should continue to engage in these appearance-normalizations but criticize others for engaging in even more meaningful ones.

Surely, any thinking person sees that there are clear benefits to the medical normalization approach. First and foremost, it sometimes works: it sometimes changes the anatomy enough that the person is not immediately recognized as being a violator of the anatomical rules of identity. Eilish Holton doesn't look conjoined any more, and it is almost certainly a lot easier, all other things being equal, to go through life a singleton than a person conjoined. Another benefit of the medical normalization approach is that, because of the prestige medicine enjoys today, engaging in a medical normalization can bring honor to a family that may otherwise be steeped in shame. If having a child born with an unusual anatomy can be construed as a temporary, fixable medical tragedy rather than as a persistent social shame, the family may find more support, financial and moral. And medical normalizations are so very concrete. They feel so certain when everything else feels uncertain. All the adults involved know their role in a surgical normalization: they must help the child through the medical procedures. When entering upon the medical normalization route, the nurses, doctors, and parents can assume the clear-cut identities of all-giving benefactors of a "sick" child; the child, meanwhile, is affirmed as an innocent who has been unjustly struck down by a whim of nature. In the medicalized quest for normality, all those involved can be liberated from the shame associated with abnormality. They can even achieve the opposite of shame, namely heroism.[24]

In the nursing literature and popular press, children who have undergone separations and other normalizing procedures are often praised as brave little fighters. For example, about one child who died following separation, a nurse said: "He remains an adored and very precious boy who holds a special place in the nurses' hearts, as they alone know the battles he has fought to achieve independent life."[25] Yet despite the comfort provided by such formulas, adults (including journalists) who become involved in separation surgeries should be cautious about ritualistically crediting the infants with heroism and bravery, as if these children were willing soldiers in a just war. Separations are not simply battles against unmitigated evils, and an infant cannot possibly choose such an undertaking. As the protagonist in the novel *Mendel's Dwarf*—a geneticist with dwarfism—notes, "In order to be brave, you've got to have a choice."[26] By attributing heroic fortitude to an infant or young child subjected to a normalization, one merely assuages the adult conscience and discourages everyone, including the child, from questioning or objecting to the procedure.[27] We must not forget that the decision maker in virtually all these cases is a person who lacks first-hand knowledge of the condition, who will not undergo the procedure, who will not suffer the costs and bear the risks. We must not forget that although normalizations *may* sometimes be wise—may even be the best choice—they are not the only option, and should not be chosen hastily.

After ten years of studying the treatment of unusual anatomies and talking with countless medical professionals, patients, and parents, I have learned that pediatric surgical procedures to normalize the appearance of psychosocially problematic anatomies are a lot more problematic than they first seem. This doesn't mean I think they should never be chosen. But I do think that, especially before any appearance-normalizing surgery is undertaken without the informed assent or consent of the patient, the following questions must be asked and their answers carefully contemplated.

What, exactly, are the goals of these surgeries? And are those goals likely to be achieved?

It is important to set aside the idea that separation surgeries are simply medically necessary procedures. Nonemergency separation surgeries almost never improve the physical health of either twin. In fact, they often leave the children's bodies—at least temporarily and often permanently—much more ill and impaired than before, and they may significantly reduce life expectancy. While they may look medically necessary in the traditional sense, they are almost always performed primarily for psychosocial reasons—because the adult decision makers believe the children will be better off psychosocially if separated, even if this means the children will lose function they would otherwise have. Sometimes, as in the case of Katie and Eilish Holton, parents choose surgery in the hopes of *increasing* some particular function, such as mobility. But in most cases, as with the Holton twins, even these "functional" issues are essentially psychosocial. Being unable to walk did not threaten Katie and Eilish's physical health. If they had never managed to walk unaided and conjoined, they might have used a wheelchair. They would have been limited by wheelchair access, but that again is a social problem, not a simple physiological one.[28]

Thus, the primary reason for most nonemergency separations is not physiological; it is not reduction of illness or increase in life expectancy.[29] It is the desire to make children who have very unusual anatomies look fairly usual, and to give two children who are supposedly trapped in a single body greater physical independence, freeing them from the stigma of conjoinment and enabling them to enjoy, among other things, a "normal" sex life. So, are separation surgeries likely to achieve these goals?

Separation surgeries do make conjoined twins look like singletons. By definition, they never fail at that (even when both patients die).[30] So they presumably reduce the degree of stigma surviving children are likely to experience. Compared to conjoined twins, children who are separated undoubtedly attract less attention from the media, from the curious, from bullies, and from gawkers. Though the patients may still visibly differ from the norm, new acquaintances presumably make fewer oppressive assumptions about them.

Does this mean they will go on to have better romantic prospects, and better sex lives? That may depend on what you mean by "better." Certainly, a twin who is separated will not, as an adult, be obliged to have sex while attached to her sibling. In other words, separation surgeries are guaranteed to satisfy singleton adults who are discomfited by the idea of a conjoined person having a sexual relationship. Whether or not separation actually improves a person's chances of finding partners for romance and sex is unclear, but it seems likely that they do in general, if not in each particular case. Nevertheless, despite the difficulties conjoined people may experience in finding sex partners and in getting married, none that we know of have chosen separation to remedy the situation.[31] Surely, many might simply be unwilling, given the choice, to risk their lives, health, or sexual sensation to a surgical procedure that might make partnering easier.

Parents are of course justified in fearing that children with unusual anatomies will grow up to be sexually rejected; even Mike and Patty Hensel, who have accepted their daughters' conjoinment, have frequently spoken of this problem. Early in the girls' lives, Mike "look[ed] toward the twins' teenage years with concern. 'They won't have the same chances as other girls,' he [said]. 'They're good-looking, and it will be tough on them.'"[32] But although the concern over children's life-long well-being is admirable, I, like the Hensels, question the likelihood that certain separations will ensure a person a better sex life. When conjoined twins who (like Abigail and Brittany Hensel) share one set of genitals are separated, they are likely to be left by the surgeries with diminished sexual function and reduced fertility. Again, there are no published studies on this; so far as I can tell, separation surgeons have not been asking their patients, years down the road, whether or not they have sexual sensation, can reach orgasm, or are fertile. But we can assume that many twins suffer diminished sexual function and reduced fertility, given the radical nature of the procedures (splitting up penises, clitorises, vaginas, uteruses, and so on), as well as reports from adults with intersex conditions who, as children, were subjected to the same invasive "reconstructive" genital surgeries.[33]

Interestingly, the same sexual norms and associated fears that motivate intersex surgeries seem to be factors in separation surgeries. Indeed, it would be hard to overestimate the degree to which anxiety about conjoined children's future sexuality motivates singletons' approval of separation surgeries. In an article praising the surgeon who in 2002 separated Maria de Jesus and Maria Teresa Quiej-Alvarez, *USA Today* reported that "*the moment* the team of more than forty doctors and nurses at UCLA Medical Center completely separated the heads of [the] one-year-old Guatemalan twin girls, neurosurgeon Jorge Lazareff paused and spoke to the assembled room: 'I said, "We now have two weddings to go to."'"[34] Obviously, the number of weddings (zero, one, or two) the girls might achieve weighed heavily on Lazareff's mind. And he is not alone. Reading the literature on conjoinment, one gets the sense that many adults seem to fear a conjoined child will grow up to be loved sexually almost as much as they fear a conjoined child will *not* grow up to be loved sexually. Regardless of which fear looms larger, sexual anxiety forms a prominent and convoluted theme in medical and media narratives of conjoinment and separation.

This concern about conjoined sexuality goes back at least 130 years, and probably much further. Recall that the anatomists who autopsied Eng and Chang Bunker specifically named the Bunkers' active sex lives as a reason to learn as much as possible about how to effect separations in the future. As respected as the Bunkers were, many people outside their circle considered it "most immoral and shocking that the two should occupy the same marital couch with the wife of one," even though one of the widows reported that "there never had been any improper relations between the wives and the brothers."[35] The men who autopsied the Bunkers argued that even if a separation of the twins had "involved great risk to life," it nonetheless "would have been well if the twins could have been separated," in view of "moral and even physical considerations."[36] I doubt it was a random choice to list physical considerations second. Disapproval of marriage—and, by implication, sexual relations—involving conjoined twins continued strong in the post-Bunker era. One of the anatomists who dissected the Bunkers, when asked to comment on the marriageability of Millie and Christina McCoy, "explained that physically there are no serious objections to the marriage of Her or Them; but morally there was a most decided one."[37] Violet Hilton was denied a marriage license specifically on the grounds of "morality and decency."[38]

Now let's consider the more general question: Do separation surgeries achieve the goal of freeing children to live independent lives as individuals? The problem with this question is that conjoined twins almost invariably state that, from their point of view, they don't need to be separated to be individuals, because they are *not* trapped or confined by their conjoinment. The question, then, is whether we ought to believe them. Given the consistency in their claims and the evidence they provide, I'm inclined to take them at their word. I therefore tend to think that, in general, separations surgeries change singletons' assumptions about particular individuals rather than liberate individuals who would otherwise feel trapped. Of course, assumptions are very important; they can become self-fulfilling prophecies. A singleton who assumes a conjoined person is incapable of independence or individuality may work against that person's efforts to achieve independence and individuality. So separation could make these efforts more successful. But again, we don't know.

Readers will have noticed by this point that uncertainty is pervasive when it comes to evidentiary questions. This is because, unfortunately, the outcomes of normalizing separation surgeries have been inadequately studied, and the few studies that have been done are flawed by persistent overgeneralization. For example, in a 1997 study of three separations of ischiopagus twins, a team from Toronto's Hospital for Sick Children found that the subjects did well in certain important respects: "All [the separated] children are independent and ambulatory with fully healed wounds. Of the six kidneys five are functioning well. Of the four children old enough to undergo continence evaluation all are fully continent and void spontaneously."[39] Yet the team's conclusion, that "conjoined twins can be successfully separated with an effective outcome," is surely too broad. The study certainly suggests that separation of ischiopagus twins often succeeds in terms of providing increased mobility and good urological function. But it tells us nothing about how the sexual sensation of the twins may have been affected by the substantial surgical changes in their genital tissue. It tells us nothing about whether these children are less likely to suffer stigma now than they would have prior to the operation. And it tells us nothing about the likely success rate of surgeries involving conjoinment of other parts of the body. The separation of, say, craniopagus (head-joined) twins is obviously a very different matter from the separation of ischiopagus (hip- or pelvis-joined) twins. So the study's sweeping conclusion is unwarranted.

A less simplistic outcome study, also published in 1997, was conducted by the pediatric surgical team at Red Cross Children's Hospital in Cape Town, South Africa, an institution known worldwide for its expertise in separation surgeries. The article looked at separation procedures in numerous types of conjoinment and went into significantly more detail than the Toronto study with regard to long-term morbidity (illness and impairment). But like all published outcome studies, it assumed—this time quite openly—that separation is a necessary and unmitigated good (though one that unfortunately sometimes results in death). The authors did not consider whether the children involved might have been better off, at least in some respects (such as mobility, fertility, brain function, sexual sensation) prior to the operation.[40] And they did not consider the data available (or collectible) on twins left conjoined. The implied question of such follow-up studies is never *whether* surgeons ought to separate in a given instance, but *how* they ought to separate.

Unfortunately, time has not improved the quality of outcome studies. In a 2002 study published in the *British Journal of Surgery*, two of the surgeons who worked on Katie and Eilish's separation, Lewis Spitz and E. M. Kiely, offer follow-up data on seventeen separations they have performed. On the surface, the article seems to offer a wealth of critical information about outcomes. But the tables show how reductive the study actually is: under "outcome" the only two possibilities provided are "dead" or "alive and well." If this is the way these surgeons are really thinking about outcomes—the patients are either "alive and well" or they're dead—there is surely a problem. Where is the detailed information about how the surviving twins are faring, in terms of morbidity? Where is the information about their psychosocial health, the reason that is given for most separations? Why are there no long-term comparisons between separated twins and conjoined twins? The simplistic conclusion that if a separated twin is alive she is necessarily well (and by implication, better off than she was) is truly disturbing.[41]

Poor quality of follow-up studies for normalization procedures persists partly because—surprisingly—most medicine is not yet evidence-based. Only now is the field of medicine beginning to favor, in decision making, well-researched aggregate data over individual clinicians' training, personal clinical experiences, and the practice of storytelling in medicine.[42] But the lack of follow-up data for most kinds of normalizing surgeries also has something to do with the peculiar nature of surgery; unlike drugs and many nonsurgical medical procedures, surgeries, at least in the United States, are largely exempt from systematic review. There is little tradition or regulation in support of rigorous systematic review. This problem is especially persistent in childhood surgeries, perhaps because of the assumption that a "fixed" child, no matter how damaged, is always better than a weird-looking or ill child; at least in such cases people *tried* to help. Mark Stringer remarks, "There's insufficient critical analysis of what we're doing, but that's partly because pediatric surgeons are so busy and sometimes don't have enough time to reflect. . . . While there has been a major drive to look at quality of life in some areas of medicine, this is an area that has been relatively neglected in pediatric surgery."[43]

There was a time when surgeons did not specialize in treating specific age groups, and so they were more apt to see the very long-term effects of their work. But today most patients stop seeing their pediatric specialists once they reach late adolescence.[44] Stringer sees this as all the more reason to do careful follow-ups: "We should be interviewing parents [and patients] years later, finding out more about what they think of the process. We should be doing psychological assessments on the children. . . . I think it imperative that we get whatever data we can . . . to know whether the doctors have failed them, or, in retrospect, whether the decisions should have been different."[45] Without this kind of data, most of the decisions being made about normalization procedures (such as nonemergency separation surgeries) are based on unfounded assumptions made by well-intentioned adults who know too little about what it is like to live with the particular condition.

Perhaps it is because there is so little substantial information about long-term outcomes that physicians seem sometimes not even to provide what little data *is* available to decision makers. I was particularly struck by this while listening to analyses of the deaths of Ladan and Laleh Bijani, craniopagus twins who choose in 2003 to be separated at age twenty-nine. Pediatric neurosurgeon Benjamin Carson of Johns Hopkins University, who helped to lead the separation team in Singapore, repeatedly told the press—and presumably the Bijani sisters, too —that there was a 50 percent chance at least one of them would be disabled or die from the surgery.[46] But as a leading expert in the field, Carson surely knew of the most comprehensive study of craniopagus separations, which had concluded that "mortality and morbidity after surgical separation of craniopagus twins are horrendous: of the 60 infants operated on, 30 died, 17 were impaired, 6 were alive but ultimate status unknown, and only 7 were apparently normal."[47] In view of this fact, it would have been more accurate to suggest that, in a craniopagus separation like this one, each twin had at a 78–88 percent chance of impairment or death. That the Bijani sisters were adults, all experts agreed, only made the odds for them worse than for infant patients: the women's skulls had thickened and hardened; their brains had developed to maturity and would therefore be less resilient. One has to wonder whether the sisters, who died from the surgery, might have chosen differently if given this information.

What do people who have grown up "uncorrected" say about living with the condition? Have some of them chosen normalization surgery for themselves, or expressed a wish that their parents had chosen it for them when they were infants?

Historical and anecdotal evidence from conjoinment and other unusual states tells us this: given the opportunity to grow up and decide for themselves, many people with unusual anatomies, including those conjoined, do not choose normalizing surgeries for themselves, at least not at a rate significantly higher than the general population chooses cosmetic surgery. Many people left to grow up with unusual anatomies report being comfortable with their bodies. They consider themselves normal, and when someone bothers to test them for psychopathology, they come up healthy at about the same rate as the general population.[48] Surprisingly often, when asked, they even say that their unusual state is *preferable*. All this should lead us to consider whether every normalization is clearly necessary for the child's mental health.

As noted earlier, in the history of conjoined twins apparently only one pair—Ladan and Laleh Bijani—have chosen separation surgery for themselves. I have been unable to locate any evidence that a conjoined person wished his or her parents had chosen separation surgery for him or her; even Laleh and Ladan Bijani apparently did not express this wish. (Ladan told reporters, "We have enjoyed being together. But we want to be together separately.")[49] Yet there are many published sources in which conjoined twins say that they accept and even prefer conjoinment to the idea of having been born a singleton.

On the other hand, there are at least anecdotal reports of a couple of older children and adults who were born conjoined and who have publicly expressed gratitude that they were separated as infants, sometimes despite having lost a sibling or suffered a physical impairment. More-over, I have found no evidence that anyone has ever expressed regret at having been subjected to separation. (By contrast, there are plenty of people who underwent intersex normalizing procedures as children and who are now angry and regretful.)

So both conjoined twins and separated twins have reported satisfac-tion with their states—though we have much more historical evidence about the opinions of the former. The singleton instinct is to believe separated twins and discount the statements of those who remain con-joined. But if we take both groups seriously, then it is unlikely a child born conjoined will grow up to regret his parents' decision regarding separation in infancy, regardless of what the parents' decide. (Of course, parents must consider more than this outcome when making decisions about separations.)

What are the minimum negative effects that will result from the surgery? And what are the maximum possible negative effects?

In the case of conjoinment the answers to these questions vary, de-pending on the nature of the conjoinment, who performs the separa-tion (including who provides the pre-op and post-op care), the way the surgeons choose to do the separation, and the health of the children. Conjoined twins who have almost two complete bodies, as Chang and Eng Bunker did, and who are separated as young infants by a capable team at an experienced medical center are unlikely to suffer many nega-tive effects. But they are in the minority. Far more often, because there is not enough tissue to go around, the twins will be left lacking body parts. When twin boys are born sharing one set of genitalia, the medical field to this day persists in giving one boy the male genitalia and making the other into a "girl" by forming female-looking genitalia out of other parts, following up with estrogen treatments, and effecting a gender identity change. One of the identical twins comes in a boy but goes out "reconstructed" as a girl. (The idea is that a boy has to have a penis to grow up as a boy, and penises are supposedly much more difficult to construct than female genitalia.)[50] This has happened in three cases, and possibly several more. At least one is reported by Spitz and Kiely, who assert without elaboration that both patients are "alive and well."[51] In another case, the child who was given the male genitalia died from the operation whereas his "sister" survived—and so the parents who had come to the hospital with two toddler-aged sons left with one daughter.[52] Lin and Win Htut were likewise toddlers when separated; surgeons gave Lin the one available penis and "reconstructed" Win as a girl, despite the fact that he had spent two and a half years as a boy conjoined to his identical twin brother (see Figure 7).[53] Apparently, such a post-separation scenario is considered less psychologically risky and more

The separation of conjoined twins — an OR nursing perspective

by Jean Savickis, RN

The arrival of 2½-year-old male conjoined tripus ischiopagus twins from Rangoon, Burma, at The Hospital for Sick Children in Toronto on July 18, 1984, presented a unique challenge to the 50 personnel who were involved in their surgical separation.

The decision to operate on the twins was confirmed on the day after their admission, after a barrage of investigative procedures. A tentative date of Saturday, July 28 was set for their surgery and the vast network of professionals to be involved in the surgery was activated. For all of the team, continuous communication became the vital link.

An initial visit with the twins allowed us an opportunity to visualize their physical limitations and the effect these would have on positioning, draping, transferring one twin to a second OR bed, etc. I also met with Dr. R. Filler, surgeon-in-chief, who outlined the proposed extent of the operation which would involve four surgical services: general surgery, urology, orthopedics and plastics.

A team of 12 operating nurses was selected, each nurse highly skilled to effectively meet the needs of the patients, anesthetists, and surgeons from each of the four services and having the ability to deal with any emergency situation that might arise. Normally, nursing coverage in our OR functions on an eight-hour shift, but as the anticipated procedure was expected to last up to 20 hours, a

Jean Savickis, is the head nurse in the Operating Room, General Surgery Service, and was the coordinator for the operative phase in the separation of the conjoined twins.

Photographs for the two articles on the conjoined twins provided courtesy of the Visual Education Department, The Hospital for Sick Children.

Lin and Win, shortly after their arrival from Rangoon, Burma.

The Canadian Nurse   December 1984   21

7.
Lin and Win Htut at two years of age, before surgical separation in 1984. They shared one set of male genitalia, and so the surgeons decided to give Lin the penis and sex-reassign Win as a girl.

conducive to good sexual functioning than a life of conjoinment. Again: there's no good evidence for this.

In my experience, the lay public tends to have a naive, overly optimistic view of normalization surgeries. This is exacerbated by the lack of qualitative follow-up data in the medical and lay literature, and the tendency to publicize full details only for the most successful separations. *Most* separation surgeries involve substantial risk, or even *necessarily* harm patients physically. After the separation of craniopagus twins Ganga and Jamuna Shrestha in 2001, the doctors made an unusual admission about what is in fact a relatively common outcome: post separation, "the two are still deformed and will suffer disabilities."[54] One medical article has estimated that only 5 percent of separated conjoined twins are ever discharged from the hospital; the rest die.[55] (Keep in mind that not all pairs forming the basis for this statistic were viable without the surgery; separations are sometimes done on an emergency basis in an attempt to save twins' lives.) Separation surgeries leave some twins brain-damaged, some bereaved of their twin (an event generally considered a significant psychological trauma when it happens to a twin born separate),[56] some temporarily or permanently movement-impaired, some without reproductive capabilities and sexual sensation they might otherwise have had, some with scars and disabilities egregious enough to cause significant social stigma in themselves, and some dead. Given the current state of our knowledge, I do not think we can rationally conclude that being separated is always better than being conjoined.

Are childhood surgeries the most appropriate or effective way to deal with the psychosocial concerns?

Again, because we do not have comparative, qualitative, long-term studies of conjoined twins and separated twins, we do not know the extent to which one group might be better off psychosocially than the other. The same is true for intersex conditions, so that a comparison of intersex surgeries and separations surgeries cannot shed light here. But what we know from reports on craniofacial surgeries is quite interesting: professional caregivers, including many surgeons, have concluded that how well the *family* is taken care of matters just as much as how well the *cleft lip* is taken care of, and may even matter more. A good cosmetic outcome may be effectively undermined by poor social support and psychological care for the family; a poor cosmetic outcome may be more effectively remedied by psychosocial interventions (including peer support) than by more surgeries.[57] This shouldn't surprise us much, since we're talking about psychosocial issues that, despite "reconstructive" surgeries, are often essentially "chronic" conditions.[58] For this reason, many institutions treating cleft lip have adopted an intensive, multidisciplinary, long-term team approach.[59] Yet this insight has been largely disregarded in other arenas of pediatric normalization, since the outcome for the *patients* has received much less attention than the outcome for their *bodies*. This mistake has been made again and again in cases of pediatric normalizing procedures.[60]

Follow-up research might well indicate that some sort of pediatric normalization procedure could improve some *parents'* mental health, in which case one could argue that it is also necessary for their children's mental health. But J. Edmund Howe has persuasively argued that, if the parents' mental health needs addressing in such cases, the parents' mental health *ought to be addressed directly.*[61] After all, in what other sort of case would doctors treat a child with medicine or surgery to improve the mental health of his parents?[62] Such a practice would likely be considered ethically questionable (if not downright unethical) in another realm, but persists in the case of unusual anatomies because medical professionals tend to see unusual anatomies as self-evident medical problems afflicting the child.

The current standard of care for children born with socially problematic anatomies does not openly or directly address the social problem of stigma and shame. In failing to do so, the standard of care—a surgical standard—inadvertently augments shame and stigma, both for the families personally and in society at large. The surgeon dealing with parents of an intersex child or of conjoined children does not say to those parents (or the public), "There is enormous social shame associated with this condition, and that is why we all feel the need to change how this child looks"—any more than the nurse dealing with a fat patient says, "I know one reason you want to lose weight is that people who are overweight are made to feel very guilty and ashamed in our society." Most medical professionals probably avoid raising the issues of anatomical shame and stigma because to do so would be to acknowledge that shame and temporarily amplify it, whereas they're trying to help get rid of the patient's feelings of shame. But they probably also don't say this because they, like most of us, tacitly accept the cultural norms that drive the shame.

Nevertheless, many parents of children with unusual anatomies and many adults who grew up with unusual anatomies have told me (and others) that because medical professionals addressed this shame only indirectly, it was actually cemented and legitimized. Surgery just confirmed that the children and their families were freaks. Obviously, physicians do not mean to sanction the social shame that comes with these conditions. But by not addressing it or questioning it, they put the weight of their cultural authority behind it. The problem that is always being fixed and attended to is the child, not the social situation. And in this focus on the child, a silent but powerful affirmation emerges: the doctor doesn't really believe that the social system is broken; it's the child that's broken.

But really, in many of these cases, the child *isn't* broken, except in terms of how others choose to form relationships around her. And so I think that in all of these cases, regardless of whether surgery is offered, professional and peer psychosocial support ought to be seriously considered—and not as an occasional add-on but as the *primary* intervention. In intersex and conjoined twinning, as in most areas of medicine, the choice not to intervene technologically—for example, with appearance-normalizing surgeries or hormone treatments—is seen by many doctors and parents as "doing nothing." But we don't have to "do nothing." Psychosocial supports are effective means of treating psychosocial distress.[63]

Debbie Hartman recalls that when she gave birth to a child with ambiguous genitalia, she begged her doctors to introduce her to parents who had been through a similar experience. "I don't care if they're in China," she remembers telling her doctors, "I want to talk with them."[64] The doctors told her that there was no one else—that this had never happened before. Even if Debbie's child had had a unique anatomy (in fact, her child had a reasonably well-known intersex condition), it would have been inaccurate to say that no other parent had ever been through anything similar. The experiences of such parents, *regardless of the specific condition in each case,* share enough features that, statistically speaking, there is *always* some other parent with experience of atypical anatomy who could be found within a reasonable geographic (or electronic) distance and to whom a new parent could be introduced. Parents report that peer support is invaluable in helping them acknowledge and sort out their complex feelings of joy, grief, fear, confusion, shame, fascination, guilt, and pride. Besides alleviating the loneliness of being the only "freak," peer support for parents (and later the children) affords access to important information and resources. Support groups also give people a forum where they can speak positively about their "defects" and organize efforts on behalf of improved social and medical care. When doctors fail to provide access to such venues, they intentionally or unintentionally keep these families in a weakened and isolated position. They harm the people they are supposed to be helping.

Could the normalization procedure be postponed until the child is old enough to make the decision?

Some kinds of normalization surgery are likely to have significantly better functional outcomes if performed in infancy. For example, the separation of craniopagus (head-joined) twins is best done before the skull has completely hardened and early in mental development. Other types are likely to have better outcomes if done well after infancy. For example, vaginoplasties (operations to construct vaginas for girls born without them) appear to provide the best long-term outcome if performed during or after puberty, rather than in infancy.[65]

As a general rule, surgeries performed early in life are less likely to cause severe scarring than surgeries done later. But functional outcome and visible scarring are not the only factors that should matter in decision making. A person's right to self-determination and to ask questions about unnecessary risk is essential as well. In other words, the question of *when* a surgery is optimally done should never supersede or erase the question of *whether* it should be done without the patient's assent or consent. Many normalization surgeries could safely be postponed until the patient is old enough to make the decision—and indeed, they *are* postponed whenever a child's physical self-determination is deemed more important than the need to provide the child with some particular kind of physical normality.[66] Waiting until a child can make the decision does not necessarily mean waiting until he or she is eighteen. Priscilla Alderson has shown that many children are able to understand and assent to surgery well before the legal age of majority.[67] In 1995 the American Academy of Pediatrics (AAP) endorsed the idea that the assent of even young children should be sought in many cases of critical medical decision making. The academy asserted not only the significance of respecting children's capabilities and perspectives, but also noted that enlisting children in the decision-making process helps them to develop as responsible moral agents. Clearly the AAP had appearance-normalizing surgeries in mind when they produced this policy; they name "surgical repair of a malformed ear in a twelve-year-old" as an example of a procedure for which consent of the parents *and* assent of the patient should be obtained.[68]

From a practical standpoint, professional psychologists and social workers can help to determine whether a given child is old enough to understand and undertake a particular decision. But also from a practical standpoint, it may be the case that normalization surgery is much less traumatic if done before the child has a fully developed familial identity and sense of self. This is because appearance-normalizing surgery is likely to be psychologically disruptive—even damaging—to the aware child and his or her family, since they may already have come to see the anomaly as acceptable or integral. Of course, such a situation—acceptance of an anomaly by the child and family—should not be seen as a poor outcome that must be prevented through surgery at the infant stage. Adults should not choose to separate conjoined twins as infants specifically because the infants and their families might grow to feel quite comfortable about the conjoinment.

In what way might the interests of the adults unduly influence their decisions?

Surgeons and parents may believe that they choose normalizing surgeries purely for the child's own good. Nevertheless, they need to examine the ways in which their own interests—what they stand to gain or lose—may shape their choices. For example, as we've already seen, parents may choose a normalizing procedure because they wish to resolve their own identity crises. And surgeons have quite a lot to gain by pursuing separations: professional and public renown, the approval of their institutions (which often benefit from the positive publicity), and the gratitude of patients' families.[69] The media swarm around separation stories, usually treating the lead surgeon as the co-star of the drama.[70] When a surgeon passes on a chance to perform a separation, colleagues at other institutions will often leap at it. (One physician I interviewed told me with obvious frustration of a recent case of this type. A surgeon, envious of a long-time rival's sudden fame following a very public separation surgery, went "shopping" around the world to find a set of conjoined twins he himself could separate, despite the fact that he and his institution had had virtually no experience in separations.) So faith in their own abilities, the admirable desire to be helpful, and concern for reputation will often incline surgeons to favor separation—as will their belief that it is not their job to seek out or consider information on the lives of people who have not had separations. In short, parents and surgeons are far from emotionally objective in the decisions they make on behalf of the children.[71]

A major constraint on adults' choices about these procedures is the lack of positive models for those who decide against surgical normalization. If we are to discourage the trend toward hyper-medicalization, we must find a way for physicians and parents who decide against unnecessary surgery to be seen as good, responsible, and caring. We must some-how alter the notion that medical intervention is necessarily the highest and best form of caring. Until we are able to do this, physicians and parents will feel pressured to choose risky, unnecessary procedures (especially if these are the only interventions for which there is insurance reimbursement). This lack of positive role models for parents has been remedied to some extent by parents like Marlene Cady and Patty and Mike Hensel, who have publicly accepted conjoinment over separation. But even in their cases, media stories tend to downplay the importance of the parents' decision, implying that the real deciding factor is the children's complicated anatomy. Meanwhile, some parents who have resisted risky normalizations have been taken to court by surgeons and charged with endangering the welfare of their children.

For surgeons, there are virtually no public models of highly respected colleagues who are hesitant about normalizations. There are, instead, many cautionary mythologies (often untraceable) about the tragedies that befell certain patients who weren't appearance-normalized.[72] Such tales may or may not be based on fact; the important point is that they can give rise to unfounded fears that impel surgeons to act. And because the birth of a child with an unusual anatomy is so widely understood as unfair—as a brutal assault by nature on an innocent child—surgeons who treat a case of "deformity" at the infant stage are virtually guaranteed to be lauded as rescuing heroes even when their patients die. Indeed, many of the men and women who go into this field do so because they see it as a noble cause to make children with anomalies look normal. In the end, surgeons may be unable to see refraining from surgical intervention as a legitimate option.[73] There are simply too many personal and institutional barriers that prevent them from considering the question of *whether* to separate, as opposed to *how* to separate.

In what other ways are the adults' choices unduly constrained?

Too often parents and doctors are constrained by a simplistic concept of parental autonomy in which the parents are understood to have free choice about their children's normalizations. Even apart from the issue of whether parents should ever be able to consent on behalf of a child in such a case, so much constrains parental choices that they can never be freely made. There are the constraints of financial resources, the constraints of self-interest, the constraints of the range of medical services offered, and the constraints of all the societal myths and biases regarding anatomies. The most prevalent myth is that an unusual anatomy must be considered a medical pathology. This is why, just after the birth of a child with an atypical anatomy, parents are often approached by surgeons rather than professional or peer counselors. They are encouraged to make decisions very quickly, even when there is no medical urgency. And if they do put off surgical normalization, they may find that although a great deal of financial support and dedication is offered for normalization, little or none is offered for psychological support. I once attended a meeting at which a psychiatrist with experience in psychosocial interventions for intersex and a surgeon advocating early "cosmetic" genital surgeries for intersex openly disagreed about the best treatment for a boy with a slightly unusual-looking penis. The surgeon, visibly annoyed, asked the psychiatrist, "How many of these psychiatric sessions will your approach take at fifty minutes each? I can fix the problem in thirty minutes!" Even *if* surgical normalization were the right choice ethically, morally, and socially, no matter how fast the surgeon's scalpel, no matter how fast the healing, some confusion, ambivalence, and doubt is likely to remain for parents and patient. Surgery is no substitute for direct treatment of the real issues.

Unfortunately, few social workers, psychiatrists, and clinical psychologists have had any formal education in treating families coping with unusual anatomies. (This, again, is a result of the belief that surgeries make the concerns of such families go away.) As a consequence, the fields of mental health and social work must develop training modules, standards of care, and subspecialties if the needs of these families are to be met (and they obviously have to be met whether or not normalizing surgeries are performed). Healthcare professionals, parents, and policymakers must also insist that mental health services be covered by insurance to the same extent that the surgeries are; at present, they often are not. The surgeon who questioned the psychiatrist about how long his approach would take was, in a way, reflecting the current insurance bias: surgeries and medications are favored over talking therapy, behavioral therapy, and social accommodation.

More radically, I would argue that because parents and surgeons are generally unfamiliar with the day-to-day lives of adults with stigmatized anatomies and because they lack a developed political consciousness about these issues, their choices about normalizations are constrained by an oppressive ignorance about the social context of their decisions. Indeed, I would go so far as to argue that families coping with unusual anatomies ought to be actively encouraged by their care providers to develop a political consciousness about difference and disability, and that the providers ought to model that consciousness. These families should be encouraged to express anger at oppressive social norms instead of at fate or themselves, and to protest, when they can, against a system that imposes shame on them although they have nothing to be ashamed about. Imposing shame can sometimes be useful and appropriate—as when someone has intentionally hurt another person unnecessarily—but at this stage in our liberal democracy, we ought not to tolerate a system that makes people who are merely anatomically different feel broken, guilty, worthless, and subhuman.

In an article entitled "Distracted by Disability," bioethicist Adrienne Asch has written: "Doctors and bioethicists shape decisions of individual patients and families, and they cannot help others make genuinely informed decisions about how to handle life with a disability if they themselves continue to be disbelieving or astonished that people with a variety of impairments can pursue life plans they find satisfying."[74] But many professionals still do not believe that a conjoined life can ever be worth living, despite so much evidence to the contrary. The bias toward separation at virtually any cost is obvious in the medical and bioethics literature. One expert judged the *survival* of unseparated twins to be "unfortunate."[75] Another concluded, in the case of newborns similar to Ruthie and Verena Cady, "It seemed unwise in view of the prolonged survivals of some dicephalus [two-headed] twins *not* to attempt separation."[76] In other words, it was the likelihood the twins might thrive that has impelled some surgeons to choose a profoundly debilitating separation. The most vivid example of this singleton bias appeared on the cover of the *AORN Journal* in 1982 (see Figure 8). The photo shows an attractive little girl who, years earlier, had undergone an emergency separation because her twin was dying. The caption reads: "Conjoined twins . . . It's rewarding when one returns to visit." Note the implications in the pronoun "one." Couldn't it be rewarding when two attached

8.

Cover of *AORN Journal*, January 1982, showing a girl who had been separated from her conjoined twin in an emergency separation. The caption, "Conjoined twins . . . It's rewarding when one returns to visit," epitomizes the singleton assumption.

siblings return to visit? I do not expect singleton parents, surgeons, and bioethicists ever to deem conjoinment as equal or superior to being separate, but the unexamined singleton bias, the fundamental lack of consideration given to the claims of those conjoined, is very troubling indeed.

Are parents fully informed when making decisions?

I think it is fair to say that parents who opt for nonemergency separations are rarely fully informed about their choices and the contexts of their choices. So what would informed consent look like? First, parents would be told how much and *how little* is known about the outcomes of the procedures offered; this would include whatever information is available on people who grow up *without* such procedures. Decision makers ought to be introduced directly to such people when feasible. (The wrong approach: showing parents scary black-and-white "freak" pictures from medical textbooks in which the subjects' anomalies are featured prominently—along with intimidating and pathologizing medical terminology—but their faces, names, and deeper life stories are not.) It would be enormously helpful if we could offer parents statistical data about outcomes both from separations and from undisturbed conjoinments—but we can't, because almost no one bothers to track cases over time. Yet even if we had such statistics, they would not replace the valuable qualitative evidence that comes from knowing individual stories.

In order for decision makers to be informed about their choices, they also need explanations—by doctors and peer counselors—of the ways in which psychosocial concerns differ from physiological concerns. They need to be educated about the evidentiary and ethical problems inherent in a simplistic surgical approach to psychosocial anatomical issues—issues which are likely to persist even if the patient's anatomy ends up looking fairly typical. And they need to know that their decisions are being shaped and limited by the choice of specialists and lay experts being offered to them. For this reason, it is especially important that they be allowed the time and resources to consider all of their options. Rushing to surgery truncates their chance to inform themselves as fully as possible.

Decision makers also need to know that no matter how good the outcome data is for the procedure being considered, there will always be a degree of uncertainty in decisions concerning whether and how to intervene. The more complicated and rare the patient's anatomy, the greater the degree of uncertainty. Valerie Miké has argued—and I would agree—that we need an ethics of evidence throughout medicine which includes a recognition of "the need to increase awareness of, and come to terms with, the extent and ultimately irreducible nature of uncertainty."[77] Too often uncertainty is used as an *excuse* for medical paternalism rather than a *critique* of medical paternalism. To seek informed consent is to make sure decisionmakers are educated about what statistical probabilities really mean for individuals.

Decision makers must also understand that the success rate of any surgical procedure is tied to who is doing the surgery and where the surgery is being done. Those assenting and consenting deserve to know that the lives and well-being of the patients may depend on which hospital and which surgeon they engage. Currently, this is a dirty little secret too often discussed only among specialists who find themselves angry about territorial incursions by competing institutions. Parents deserve to know how critical their choice of institution and surgeon may turn out to be.

Is it ethical to choose a procedure aimed at radically altering a child's anatomy for psychosocial reasons only, when the surgery will certainly undo the body the child was born with?

Some people might think that as diagnostic and surgical techniques get better, ethical questions about childhood appearance-normalizing surgeries will evaporate. But this is far from true. We will always have to ask why we are choosing to "fix" a body that the individual may ultimately find good, and even superior to the norm. So are surgical normalizations performed for psychosocial reasons morally flawed because they seek to fix a child who is not broken? On most days, I don't think so. It seems to me, as a parent and a normate as well as a historian, that an unusual anatomy might lead to an *unnecessarily* painful psycho-social existence, particularly if it is a very obvious difference that could be remedied with little risk to a child's health, functioning, and sense of self. But I do think we are too quick to let the chips fall in favor of normalizations, believing they are always the best option or resigning ourselves to seeing them as the best we can hope for.

3

What Sacrifice

The vast majority of surgical separations performed on conjoined twins aim at saving both children, but sometimes there just aren't enough vital organs to create two viable bodies. When faced with the likelihood that both children will die soon if left intact but that one might live if separated, surgeons and parents sometimes opt for a "sacrifice surgery." Doctors effect the death of one twin, separating one head and other duplicated parts from vital organs such as the one working heart, in an effort to construct one viable child. Sacrifice separations are very different from emergency separations performed immediately after the death of one twin. In the latter instance, the tissue on that side of the body is already in a state of decay and surgeons are trying to save the remaining twin, who will die if left attached. In sacrifice separations all of the tissue is functioning until surgeons cut off one twin from life-sustaining organs. During and after the procedure, surgeons try to salvage from the sacrificed twin whatever parts are needed to reconstruct the other sibling. For example, in one such operation performed in Arkansas in 1977, surgeons (unsuccessfully) attempted to transplant an arm from the sacrificed twin to the twin they were trying to save.[1] The sacrificed twin is thus treated as a donor of its share of midline organs, of skin, bones, and any other part that the surgeons need for their reconstructive work.

According to public records, the first sacrifice of a conscious twin occurred in the United States in 1955. Since then, there have been at least eleven sacrifice surgeries performed worldwide.[2] (The number may be significantly higher. Not all separations are recorded publicly.) Sacrifice surgery has sometimes been described as "killing one child to save the other," and while surgeons often understandably object to the term "killing," they are sharply aware of the procedure's moral implications. The first surgeons to perform the operation sought assurance from law enforcement officials in advance that they would not be prosecuted for homicide,[3] and parents and medical professionals sought assurance from religious advisors that their participation was not immoral.[4] Today approval is typically sought from internal hospital ethics committees,[5] and although ethics committees apparently have never tried to forestall a sacrifice surgery, surgeons who perform the operation invariably report feelings of intense emotion at the moment of sacrifice. One surgeon involved with the sacrifice of Amy Lakeberg in 1993 "said that when the blood vessels connecting Amy with Angela were severed, cutting off the blood flow and causing [Amy's] death, 'nothing was said, but I know everybody felt it.'"[6] In the sacrifice of Rosie Attard in 2000, "two paediatric surgeons, Alan Dickson and Adrian Bianchi, elected to make the final cut together as they felt it was inappropriate for one person to shoulder the burden of consigning [Rosie] to death."[7] Dickson recalls, "It was a very intense moment. We looked at each other because we knew what we were doing at the time. The [surgical] theatre was very quiet. People knew what was happening and it was done with great respect. . . . It was a shared experience which I have to say I didn't relish."[8]

Medical professionals involved in sacrifice surgeries seek psychological comfort by rationalizing that "the inherent anatomy is the determining factor" in the decision to end the life of one twin.[9] In other words, *they* aren't making the decision; *nature* is. The team that sacrificed Amy Lakeberg to save her sister, Angela, subscribed to this way of thinking when they claimed that "the operation itself does not actually determine which twin will survive; rather, this is determined by the twins' condition before surgery even commences."[10] But in spite of this reasoning, obviously the doctors *do* decide that one child will die during the surgery, and they are the ones performing what amounts to asphyxiation. As bioethicist George Annas has remarked, the belief that "objective medical criteria . . . decide" that one twin will die and one will be saved constitutes a "fiction that the decision is 'an act of God.'"[11] The decision is clearly an act of men and women—doctors, parents, judges—and not of nature, God, or the children themselves. It is a decision stemming from the beneficent belief that "it is preferable to intervene to save one life, if possible, rather than to permit the inevitable loss of both."[12]

Certainly, little emotional anguish accompanies the sacrifice of some conjoined twins—namely, those that appear to be vestiges of an almost-twin, mere excrescences of tissue lacking anything remotely resembling consciousness. Known in the medical literature as "parasitic" twins, these can range from extra limbs to a formless mass.[13] Biomedical experts theorize that "parasitic twins result from embryonic death of one twin, leaving various portions of the body vascularized by the surviving" twin.[14] Virtually any body part can be replicated and attached—legs, arms, torsos—and the conformations of parasitic twins can be quite surprising.[15] Parasitic twinning may cause appreciable psychological or physiological stress for the conscious twin (who is known in medical lingo as the "autosite"), and it is not hard to see from illustrations of radically asymmetrical twinning why removal of the nonconscious "parasite" is generally considered uncontroversial (we saw an instance of this in Chapter 1, Figure 4).

But cases in which surgeons intentionally end the life of a conscious twin—however good the intentions—*are* morally especially problematic and worth examining closely, in part because sacrifice surgeries carry implications for other practices, including passive and active euthanasia, vital organ donation, and surrogate decision making. Should doctors be allowed to hasten the death of one person when they are trying to save another? Can twins who share vital organs be regarded as something less than two persons, and are they exempt from the usual prohibitions on killing? Who ought to decide how and when a child will die and how a child's organs will be distributed? Should people born with unusual anatomies be subject to a different ethical framework from the one that protects people born with more typical ones?

There are three sacrifice separations for which we have a good deal of information: that of Amy and Angela Lakeberg, born in Chicago, June 29, 1993, and separated in Philadelphia, August 20, 1993; that of Darielis Milagro and Sandra Ivellise Soto, born and separated in Boston on May 30, 1999; and that of Rosie and Gracie Attard, born in Manchester, England, August 8, 2000, and separated there November 6–7, 2000. Considering these cases together will help us to see certain patterns—in medical practice, public reaction, and ethical reasoning—as well as to appreciate just how hard it can be to sort out why some sacrifice surgeries may seem right, and others wrong.

Sacrifice surgeries had been taking place sporadically and quietly for about four decades before the 1993 surgery that sacrificed Amy Lakeberg, the first separation to spark widespread public discussion about the ethics of such procedures. When Reitha ("Joey") Lakeberg became pregnant, she and her husband, Kenneth, were living in Indiana near their extended family. Early in the pregnancy, sonograms revealed that Joey was carrying conjoined twins, and the couple decided to seek care at Loyola University Chicago Medical Center. From the start, the prognosis for the twins was grim. After about seventeen weeks, Joey and Ken knew that their daughters Amy and Angela would come into the world "joined breast to belly, with a fused liver and a shared heart," and that the heart was malformed, containing six chambers instead of four.[16] Nevertheless, they decided against abortion. According to a 1993 news report, "Although the Lakebergs are Roman Catholic, Reitha says her decision was only 'partly' because of her religion. 'It's just the way I felt.'"[17] Joey, young and healthy, carried the pregnancy to thirty-seven weeks, at which point Amy and Angela were born by cesarean section. Despite normal brain readings and other hopeful signs, "the babies became ventilator dependent by six hours of life. The way in which the chest was joined did not allow full inspiration and expiration, resulting in respiratory failure."[18] And so Loyola's ethics team held a meeting to try to figure out what to do. As is standard practice for this sort of consultation, "the family was not present at this meeting, although they knew it was taking place and were told immediately thereafter about the recommendations."[19]

The Loyola ethics committee knew that because the girls had only one heart, separating them would entail the death of one twin. One of the girls would have to be sacrificed. But the fact that the separation would involve killing one twin was not the reason the committee recommended against it, contrary to what one might assume in view of Loyola's Roman Catholic affiliation. In fact, the team believed that "the rule against killing" could have been "suspended" if "the likelihood of saving one twin by killing the other [had been] much higher than it was."[20] They recommended against separation because, given what they knew of previous similar cases and the features of Amy and Angela's anatomy, they had little reason to believe that the saved twin would do well even after separation. The twins' heart was so malformed that the sacrifice separation did not seem worth the trauma it would inflict on both children and the substantial cost it would impose on the hospital and the healthcare system. The committee counseled against separation: "Social justice demanded that resources, time, and professional energy be devoted to better outcomes, especially in a time of health care rationing through managed care and competition."[21] Later, the neonatologist who attended the children at Loyola also conceded that the recommendations were shaped not only by their "dismal chances for long-term survival" but also by "other factors, such as the twins being ventilator dependent and having a dysfunctional family with a suboptimal home environment."[22]

Despite being encouraged by the Loyola physicians to do so, Joey and Ken were not interested in taking Amy and Angela off life support and letting them die naturally. They were unpersuaded by concerns about costs to the children and to society. And so, apparently driven by desperate hope as well as a transparent hunger for attention—the neonatologist working with them later said "the family was essentially addicted" to the media and so their "'go for the glory and defy all odds' mentality prevailed"[23]—Joey and Ken sought a team of surgeons willing to perform a separation on their daughters. They found such a team at Children's Hospital of Philadelphia, an institution known by the unfortunate acronym "CHoP." But even the CHoP specialists, led by Dr. James O'Neill, Jr., were hesitant, understanding the gravity of the girls' condition and knowing that no previous surgery of this type had succeeded. Up to that point, the longest-surviving subject of a sacrifice surgery had lived only three months, and a fairly miserable three months at that. The doctors at Loyola had estimated the likelihood of survival for a single Lakeberg twin at less than 1 percent. The CHoP team estimated there was a one-in-four chance that one twin would make it through the surgery, but were unsure of the chances for long-term survival.[24] The Lakebergs, however, thought they might "win the lottery" on this one and save one daughter.[25] The surgeons at CHoP concluded: "It was the parents' right to decide whether to proceed with the surgery. . . . The physicians would not take this right away from them."[26]

The surgery took place on August 20, 1993, when the girls were about six weeks old. By this time, the physicians and the parents knew that the twins "had distinct and different personalities, and had separate heads, arms, legs, spinal columns, and kidneys."[27] Philosophers enjoy pondering the question of whether conjoined twins are two people or one,[28] but no one who knew Amy and Angela thought of them as anything but two children. Amy—known to her parents as "the ornery one," yet considered the physically weaker twin—was chosen to die in the separation procedure. Doctors thought that Angela's anatomy gave her a slightly better shot at survival. Just before the operation, nurses painted Angela's fingernails pink, in an effort to prevent the surgeons from being confused about which "half" was to be salvaged: they were to save the girl with the pink fingernails. Amy's bare fingernails signaled her fate. She died about two-thirds of the way through the procedure,[29] and a few days later was buried in a small white casket in Roselawn, Indiana.

Angela survived for about ten months after the operation. Reports differ as to how well she was doing when she died unexpectedly of pneumonia on June 9, 1994, a couple of weeks short of her first birthday. Initial news reports claimed that "Angela's brief life was largely free of suffering. Repairs to her heart had rendered it fully functional. Her chest was somewhat misshapen" because her once-shared heart distended the chest wall, but her wounds were "healing well. Angela did not spend her days entangled in tubes and wires. She needed no sedatives or painkillers or emergency trips to the operating room."[30] But later reports pointed out that Angela had "remained ventilator dependent and . . . hospitalized her entire life."[31] It was true she had had a chance to taste food by mouth near the end of her life, but because of her breathing problems she was still being fed by tube. And though her doctors, nurses, and therapists had been trying to wean her from the respirator, she spent much of her time up to her neck in a negative-pressure device (a machine similar to an iron lung) with her head immobilized. This confinement "prevented normal development."[32] She never left the cardiac intensive care unit of CHoP, living flat on her back, sometimes craning her head to watch Disney videos, much of the time being tended by nurses, doctors, and therapists.

There is no doubt Angela Lakeberg was the subject of much loving attention at CHoP. "Because she was at the hospital so long and because of her sweet, compliant personality, she was much beloved by the staff," said her nurses.[33] Occasionally they would remove her from the ventilator and give her a bath, which she loved: "She was smiling and laughing, and we'd dry her off and hold her a little while, just to get some human contact," before she went back into the respiratory-aid machine.[34] After Angela's death, one nurse remembered: "Even the trash supervisor [who] comes around in mid-afternoon—she'd stay for twenty minutes" to interact with Angela.[35]

By contrast, Angela's parents were largely absent from her life. Joey managed to visit her only three times after the separation surgery, and held her only once. (Reportedly, financial problems and obligations in Indiana made trips to Philadelphia difficult for Joey.) Ken saw her even less. A singularly unsympathetic character, he spent much of Angela's brief life in trouble with the law. At one point he was charged with misusing funds donated for the children's care.[36] The doctors and nurses who had put so much effort into saving Angela grieved for her when she died. The director of the critical-care unit, Dr. Russell C. Raphaely, told the *New York Times*, "Angela Lakeberg was a sweet little girl. . . . We're quite sad. All of us kind of consider ourselves surrogate parents for Angela."[37] Particularly painful was the fact that just before her death they had begun to think she would eventually be able to go home. She would have been seriously disabled and would probably have always required respiratory aids, but she would have been out of the hospital and living with her family. Instead, she went home to Indiana in a casket, to be buried next to her sister.

The Lakeberg separation occurred at the start of the first Clinton administration, when the nation was embroiled in a debate over spiraling healthcare costs. As a consequence, most of the ethical discussions about the case centered not on the ethics of "killing one to save the other," but rather on the question of whether it made sense to spend so much money on a single patient with such a poor prognosis. Angela's care cost more than a million dollars, probably almost one and a half million. Her parents were uninsured, so Indiana Medicaid ended up paying about six hundred thousand dollars for post-surgical care; the hospitals chipped in the rest.[38] Objections to the cost were raised even after the surgery, when Angela's prognosis seemed relatively good. Numerous commentators—religious, legal, political, and medical—challenged or dismissed outright the claim that no one ought to impose financial limits on heroic medical procedures aimed at saving very sick children. Ronald Dworkin, professor of jurisprudence at Oxford University and professor of law at New York University Law School, wrote in the *International Herald Tribune*: "Any nation that tried to provide every possible treatment, no matter how expensive, even when the treatment had only a small chance of working, would have little money left for obviously valuable medical policies, like immunization for children, or for education, or for sustaining an economy so that its people could have rewarding jobs. . . . So the rescue principle, noble as it seems, must be abandoned." In Dworkin's opinion, it was not only Amy Lakeberg that was sacrificed at CHoP: "True respect for human life was sacrificed, in Philadelphia, to bad slogans about sanctity and rescue, and, perhaps, to a dangerous love for heroic medicine for its own sake."[39]

Dworkin had plenty of company in his views. A number of prominent Christians, such as Stephen Lammers, coeditor of *On Moral Medicine* and *Theological Voices in Medical Ethics,* considered the situation from a theological and religious standpoint and came to the same conclusion: that spending huge sums of money on one very ill child was inappropriate at best. "It is grimly ironic," said Lammers, "that we live in a society that permits a procedure such as the one performed in the Lakeberg case, but at the same time permits . . . its infant-mortality rate to be one of the highest in the industrialized world. Such a state of affairs cannot be defended on any grounds."[40] Academic ethicists concurred. Arthur Caplan, a professor at the University of Pennsylvania and a government advisor, summarized the feelings of many when he told the *Philadelphia Daily News* the day after Angela's death: "Am I saying that rich kids and people who can fund-raise are the only ones who should have the Lakeberg operation? Yes. . . . This is not something that should be supported with public money."[41]

Dissent came mostly from pediatric specialists who were horrified at the idea that money might be the only thing standing in the way of saving a child like Angela. Dr. Jonathan Muraskas, the neonatologist who first cared for Amy and Angela at Loyola, was, like many of his colleagues, unable "to put a price tag on human life."[42] Meanwhile, Dr. Alan Fleischman of Albert Einstein College of Medicine in New York conceded that an organized and "broad-based consensus" about healthcare expenditures might in the future limit funding for procedures like the Lakeberg separation, but insisted that such decisions could not fairly be made prior to such a consensus simply by deciding on a case-by-case basis. "We must not allow society to choose the smallest and most vulnerable citizens to be the first to sacrifice for the good of others," declared Fleischman. "We should not ask physicians to ration at the bedside and become adversaries of, rather than advocates for, their patients."[43] Fleischman's choice of language is unintentionally ironic, because of course the surgeons who performed the separation *did* choose a small and vulnerable citizen (Amy) to sacrifice for the good of another (Angela). They *did,* quite literally, ration at the bedside. But Fleischman intended his comments to apply only to money, not to flesh, because for him (as for nearly every other commentator on the case) questions about the financial costs obscured questions about choosing to kill one sibling to save another. In his defense of the money spent on saving Angela, Fleischman asked rhetorically: "On the basis of the medical facts of the Lakeberg case, was the surgery so different from that offered in the face of other major congenital anomalies in singleton infants?"[44] Surely it *was* quite different. It involved deliberately ending the life of a mentally alert child.

In terms of ink spilled and hands wrung over the ethical problems raised by the Lakeberg case, the issue of the actual sacrifice of Amy was not even the runner-up. The second most frequently asked question was whether the separation surgery was experimental rather than therapeutic. Should Joey and Ken ever have been led to believe that the surgery might help one of their daughters achieve a good life? A *Newsweek* article presented the views of John La Puma, an ethicist at Lutheran General Hospital in Chicago: "Parents like the Lakebergs should be told that the surgery their babies would undergo is a research experiment, not a treatment. . . . 'It shouldn't be portrayed as being for the babies' good.'"[45] Many agreed with this opinion because no sacrifice surgery had yet been successful, and because the profound malformations from which the girls' suffered (particularly the six-chambered heart) could not be solved by sacrifice surgery. Lammers went further, maintaining that the surgery did not even reach the threshold of experimental medicine: "It was not even an experiment in the sense that we could reasonably expect to learn something that could help other children, as we might, for example, with experiments performed on children with leukemia. This experiment had little chance of helping other children and very little chance of helping the survivor."[46]

These sorts of questions about costs and benefits were and are worth asking. And we must understand how intense the national debates over healthcare reform and experimental treatments were in 1993 and 1994. But I still find myself, as a historian, looking back at the Lakeberg separation and wondering why questions about financial cost and experimental medicine dominated so clearly over questions about the ethics of "killing one to save the other." There was a curious silence on this topic.[47] One can only suppose that the tacit way of dealing with the issue of sacrifice surgery was what ethicist George Annas describes (and decries) as the "monster approach"—an approach based on the assumption that conjoined twins "are so grotesque that they are not really human. Therefore, we are justified in doing anything medically reasonable to make at least one of them 'human,' even if it will very likely result in both of their deaths."[48] It is unthinkable that the Loyola ethics committee would have considered suspending the "rule of killing" if the twins had been born foundering twin singletons and vital-organ donation by one to the other might have meant one could be saved. Where conjoined twins are concerned, the normal rules don't seem to apply. But why not?

I fully understand the instinct to make every possible effort to try to save one child when parents are facing the prospect of losing two. How awful it must be for parents to be looking forward to the birth of two babies and then learn that both will probably die very young. Yet no matter how justified the ends, it is troubling to see surgeons actively cause the death of a child like Amy—who was obviously conscious and as entitled to the conjoined heart as her sister. The fact that the surgeons found the twins hard to tell apart (so hard that they had to use nail polish to label the slightly more viable twin), the fact that the doctors and nurses found the sacrifice disturbing, the fact that Amy and Angela together were continuing to progress developmentally, the fact that other children like them had lived well beyond the life spans doctors had predicted for them—all this ought to give us pause. A long pause. Amy Lakeberg's sacrifice looks uncomfortably like a heart-and-liver "donation" from a person with active brainwaves. Surely this issue needs unpacking.

Yet the only publication I can find that presents an in-depth analysis of the sacrifice was written in 1996 by some of the people who were directly involved, including one of the ethicists from Loyola, David C. Thomasma, the original attending neonatologist at Loyola, Jonathan Muraskas, and the surgeon who performed the sacrifice surgery at CHoP, James A. O'Neill, Jr. Along with three colleagues, they produced an admirably critical and obviously heartfelt examination of the Lakeberg case for the *Hastings Center Report,* a leading journal of medical ethics. The authors declared that "from the standpoint of human rights, both twins had a right to life" and that one cannot circumvent this problem by claiming that Amy, the physically weaker twin, was a mere "appendage" or "parasitic" twin.[49] "The only way to justify taking the life of one in favor of another would be through surrogacy and presumed consent." Such reasoning "holds that, should one of the twins be able to speak, she might voluntarily give her life so that the other twin might have even a very slight chance for survival, given the fact that both will certainly die if nothing is done."[50] When children cannot speak for themselves, the parents, as surrogates, may make a "presumed consent" choice on their behalf.

But this line of thinking has several obvious flaws. First, there is the question of whether anyone should be able to make such a grave choice on behalf of someone else. There might be some circumstances in which consent ought never to be presumed. Suppose Angela Lakeberg had survived and eventually learned that her identical twin sister, Amy, had been killed because people believed that Amy and Angela would have wanted this. What emotional trauma would Angela have had to face? Would she have been obligated to believe she wanted her sister to die? Should we assume that Angela would have traded her sister's life for her own?

Second, as noted in the introduction, there has been only one case in which twins old enough to make a decision have sought surgical separation. And there has never been a case in which a twin has agreed to risk or lose his or her life for the sake of a conjoined sibling. I cannot even find a case in which, following the death of one twin, the surviving twin specifically asked to be separated from the dead sibling. How can one automatically assume that if Angela and Amy had survived to an age at which they were capable of making decisions, they would have differed in their attitudes from their cohort?

Third and perhaps most important, if the girls had been singletons, it is inconceivable that physicians would have considered actively causing the death of Amy to save Angela. There is no other realm of medical care in which doctors can legally and ethically use a mentally functioning person as a vital-organ donor. Even if the consent of the potential organ donor were explicitly obtained, such a procedure would not be executed. And infants certainly cannot give their consent.

Having talked with and read about people who are conjoined, and having studied many of their personal histories in depth, I can see no reason for treating them according to a unique moral framework simply by virtue of their conjoinment. One possible response is to say (as one surgeon did in conversation with me) that it wasn't as if you were taking the heart from Amy and giving it to Angela. But how *wasn't* it like that?

Seven years after the Lakeberg separation, questions about the ethics of "killing one to save the other" did come to the fore, in another context. The case of Rosie and Gracie Attard was marked by an intense battle: the doctors wanted to act to save Gracie, but the parents were unwilling to see Rosie sacrificed in the process.[51] The doctors believed that sacrifice surgery was the moral course of action—that *not* sacrificing Rosie would be equivalent to causing the death of Gracie—and so they took the Attards to court, suing for the right to proceed with the surgery. Over the course of several months, the physicians won two legal rounds, and the parents finally gave up. The girls were separated, Rosie was sacrificed, and Gracie was (and still is being) reconstructed with surgeries and rehabilitative therapies.[52]

Michaelangelo and Rina Attard lived on the Maltese island of Gozo and had been married about a year when Rina became pregnant. Four months into the pregnancy, ultrasound revealed to Rina's obstetrician, John Mamu, that she was carrying conjoined twins. The girls were joined at the pelvis, with the lower ends of their spines fused and their spinal cords joined. Their heads were pointing in opposite directions, their legs jutted from their torso at sharp right angles, and their genitals were located on the sides of the body, between the legs (see Figure 3F, in Chapter 1). Obviously this is a complex and challenging form of conjoinment, even if all else about the babies is healthy. In consultation with Dr. Adrian Bianchi, who was working part time in Malta, Dr. Mamu advised Michaelangelo and Rina to go to Bianchi's home institution—St. Mary's hospital in Manchester, England—to seek the best care. Rina and Michaelangelo did so, with the belief that the surgeons at St. Mary's might be able to save both girls.

Physicians at St. Mary's advised Rina that her daughters might not make it to term. They recommended that she give birth early by cesarean section, but Rina, "afraid of what the birth would bring,"[53] refused the advice. She later told a reporter, in her broken English, "To tell you the truth, this was a little bit terrif[ying] for me, and so I wanted that to take longer and longer so, like, it never happens."[54] She gave birth on August 8, 2000, two weeks overdue. Though Gracie and Rosie looked somewhat symmetrical externally, they had developed quite asymmetrically. The only midline shared organ was a large urinary bladder. Each girl had a brain, heart, lungs, liver, and kidneys. But Rosie's brain seemed quite underdeveloped. Doctors described it as a "primitive" brain; it lacked much of the growth and function considered normal in a newborn. Her head and neck were swollen and showed signs of oxygen deprivation. Though she eventually opened one eye and moved her limbs, she was not responsive in the way one expects of a conscious child. Gracie, on the other hand, was alert and conscious from the moment of birth.[55] She cried and responded to stimulation. The difference in their brains would not, of course, have been the limiting factor in a separation attempt to save both. It was the difference in their hearts that would make the separation a sacrifice surgery: the heart on Gracie's side—Gracie's heart—was keeping both girls alive. Rosie's heart was enlarged and abnormally formed, incapable of sustaining life. Her lung tissue was also poorly developed and inadequate, meaning that Rosie was relying on Gracie for pulmonary support as well.

The staff at St. Mary's explained to the Attards what they wanted to do: separate the twins and save Gracie. Given the girls' anatomy and the signs of stress on Gracie's heart, the doctors were convinced that if they did not separate the girls soon, both would die, probably in three to six months.[56] There was a good chance they could save Gracie through a surgery which would cut Rosie off from the heart and lungs keeping her alive. Speaking in favor of the sacrifice surgery, Harry Applebaum, a Los Angeles surgeon who had previously performed a separation surgery, told the BBC, "I think in this situation we do have a person [i.e., Gracie] who can function normally following an operation. On the other hand, we have a person [i.e., Rosie] who probably never will function normally. I think most doctors I am aware of would try to salvage somebody who can have a fairly normal life."[57]

But Michaelangelo and Rina, appealing to their Roman Catholicism, felt that the act of separating Rosie from Gracie would be tantamount to murder. They saw both Rosie and Gracie as living daughters, and could not sanction the idea of killing either one, even if this meant that one would be "salvaged." Michaelangelo and Rina believed both girls "should be allowed to die, 'if that is the will of God.'"[58] Rina later recalled that her feelings were bolstered not just by her faith, but by her growing relationship with both girls: "The more I stay with them, the more I do things for them, the more my love grows for them."[59]

Frustrated with the parents' willingness to let both children die when one could apparently be saved, the doctors took the Attards to court, in England. The suit was brought by the surgeons at St. Mary's as well as by the Central Manchester Healthcare Trust, and according to their counsel it was brought not "to authorise an unlawful action," but rather "to ask whether or not the operation which we believe to be in the best interests of the twins is in fact lawful. . . . Our submission is the operation would be lawful."[60] The physicians pursued the sacrifice because they were deeply disturbed at the thought of Gracie dying—or living for a time anatomically bonded to her underdeveloped sister—when she had every chance, in their opinion, of living a normal life if separated.

At the court hearing, Michaelangelo Attard tried to tell the judge what he felt, but he broke down in the midst of doing so, and the court had to accept the parents' written testimony. The document reads: "We cannot begin to accept or contemplate that one of our children should die to enable the other one to survive. That is not God's will. Everyone has the right to life, so why should we kill one of our daughters to enable the other one to survive. That is not what we want and that is what we have told the doctors."[61]

What was missing in most of the news accounts and the subsequent analyses (but not entirely from the judge's considerations) was an account of the other source of Rina and Michaelangelo's reluctance to approve the separation: besides being unwilling to kill Rosie, they felt deep distress at the thought of having to raise Gracie, who was likely to be seriously disabled after separation. Pro-life and disability rights groups— ostensibly on the side of the Attards in objecting to the sacrifice surgery—conveniently ignored the fact that the Attards seemed to be seeking the natural death of both daughters in part because they didn't think they could deal with the challenges posed by a seriously disabled child. The Attards knew that their homeland was a place "with very few, if any, facilities . . . to cope with a disabled child,"[62] and a place where disability equated with shame. The judge noted in his ruling that midwives who had worked with the Attards in Malta worried that if the Attards returned home with a seriously disabled Gracie, people there would assume the parents had committed some terrible sin and the family would be treated badly.[63] With the bleak environment for a disabled child that Malta seemed to offer, Rina and Michaelangelo feared they would end up having to give up Gracie to a foster family in England, where she could get the care she would need. But they did not want to be put in a situation where they would feel the need to give up their daughter to foster parents.

Justice Johnson was responsible for adjudicating this thorny dilemma. On August 25, 2000, he ruled that the surgery could legally proceed. In his decision, he drew a stark contrast between the twins, employing the physicians' analysis of the situation but also recognizing the parents' distress and revealing what he himself valued in a life and a future:

[Gracie] is a bright alert baby, sparkling, sucking on her dummy, moving her arms as babies do and, in the words of one of the doctors, "very much a with-it sort of baby." She has a functioning heart and lungs. Her legs are set wide apart but that can be rectified by surgery and the probability, so I find, is that separated from [Rosie], [Gracie] would be able to lead a relatively normal life, probably walking unaided, probably attending school and probably being able to have children. However, there might well be problems for her physically, including double incontinence, but this and the other problems would be capable of solution, including perhaps by surgery. . . . For [Rosie], things are very different. Her face is deformed, but more importantly she has no effective heart or lung function. She lives only because of her physical attachment to [Gracie].[64]

Despite his belief that the children would die within six months if left attached, Justice Johnson tried to imagine what it would be like if the children defied expectations and survived conjoined. He envisioned Rosie "in pain but not able to cry. One very experienced doctor said she thought it was an horrendous scenario, as she put it, being dragged around [by Gracie once she learned how to crawl] and not being able to do anything about it. . . . [Rosie's] life, if [she were] not separated from her twin, would not simply be worth nothing to her; it would be hurtful . . . to prolong [Rosie's] life, for those few months would, in my judgement, be very seriously to her disadvantage." Justice Johnson concluded that the sacrifice would not only be in Gracie's best interest; it would also be in *Rosie's* best interest, because it would kill her. In Johnson's view, Rosie's life as it was, and as it might become, would be worse than death. Before concluding his analysis of the children's best interests, Justice Johnson repeated his optimistic belief that medicine would "cure" any "social and emotional" as well as physical problems Gracie might face after surgery.[65] He felt fairly certain she would come through it a "normal" child.

How could he justify his decision to allow a separation that would kill Rosie? Justice Johnson asserted that "the court will never authorise any step actively to terminate life, even to relieve misery and even if the patient or a parent so consents." But, he noted, "withdrawal of treatment, including even the withdrawal of feeding," was permissible.[66] If one could imagine the clamping off of the blood supply flowing from Gracie to Rosie as the clamping off of feeding, then the sacrifice surgery would be a permissible act. And Johnson could so imagine it. He concluded that the sacrifice surgery constituted a legal act of "passive euthanasia in which [Rosie's] food and hydration would be withdrawn (by clamping off her blood supply from [Gracie])."[67]

By this time the story was attracting attention all over the world as an unprecedented battle. In the typical news account, on one side were the third-world, Roman Catholic, pro-life Luddite parents who wanted to let their babies die rather than see one killed intentionally. On the other side were the modern, high-tech British doctors who wanted to sacrifice one defect-ridden child to try to salvage a decent life for her normal (and trapped) sister, who would otherwise be unjustly dragged into death. Intensifying the religious aspect of the controversy, on August 28, 2000, the Vatican offered the parents a safe haven at an Italian hospital if they wanted to remove their children, against medical advice, from St. Mary's.[68] Instead, Rina and Michaelangelo chose to appeal Justice Johnson's ruling and the case went to a three-judge panel. The panel sought advice from legal experts in Australia, South Africa, and Canada, but to no avail: no similar case had been heard elsewhere.[69]

In the appeals process, the position of the doctors, like Justice Johnson's, was "that the operation could be seen in the same way as the removal from [Rosie] of a life support machine, and her death [therefore] considered a natural consequence of her condition."[70] But the lawyers for Rina and Michaelangelo Attard reiterated the parents' opinion: that separating the girls would be an act against God's will. They also reiterated the Attards' concerns about raising a disabled Gracie in Malta (the parents were much more pessimistic than the doctors about her chances for a normal life) and challenged Justice Johnson's and the doctors' claims that Rosie's situation was "demonstrably intolerable." The Attards did not believe that "immediate death was in her best interest."[71] Archbishop Cormac Murphy-O'Conner, head of the Roman Catholic Church in England and Wales, agreed with the parents. In his written

testimony, he warned the three judges that "'a very dangerous precedent' would be set if the judges ruled that it could be lawful to kill a person 'so that good may come of it.'"[72] He insisted they recognize first and
foremost the sanctity of every life.

After agonizing over the claims, counterclaims, legal precedents (or
lack thereof), and evidence, the three judges all ruled in favor of the le-
gality of the surgery. But each gave somewhat different reasons. Lord
Justice Ward agreed with Justice Johnson that the operation was surely
in the best interests of Gracie. Like Johnson and the doctors, he felt
quite optimistic that doctors would over time be able to give Gracie a
near-normal body. Perhaps she would have an uneven gait; perhaps she
would need to wear a colostomy bag. Still, he thought she was likely to
walk, likely to have bladder control, likely to have a functional vagina
and therefore a normal sex life. But Ward, unlike Johnson, did not think
that the separation was in Rosie's best interest. "The operation has . . . to
be seen as an act of invasion of [Rosie's] bodily integrity, and unless
consent or approval is given for it, it constitutes an unlawful assault
upon her." Ward maintained that the act of sacrificing Rosie could not
be seen as a passive act—an act of omission, like a withdrawal of artificial life support. In his opinion, the act of sacrifice was an act that
might, if not otherwise excused, be likened to murder.

What, then, did Ward see as the legal justification for this act? Pointing to the question of "right to life," he asked why Gracie's right to life
should not be considered as seriously as Rosie's. After all, by keeping
Rosie alive through conjoinment, the doctors could be said to be inter-
fering with Gracie's right to life. Was Gracie not entitled to a reasonable
shot at self-defense?

> [Rosie] may have a right to life, but she has little right to be alive. She is
> alive because and only because, to put it bluntly, but nonetheless accu-
> rately, she sucks the lifeblood of [Gracie] and she sucks the lifeblood out
> of [Gracie]. She will survive only so long as [Gracie] survives. [Gracie]
> will not survive long because constitutionally she will not be able to cope.
> [Rosie's] parasitic living will be the cause of [Gracie] ceasing to live. If

[Gracie] could speak, she would surely protest, "Stop it, [Rosie], you're killing me." [Rosie] would have no answer to that.[73]

So the sacrifice could be performed legally because it constituted "quasi self-defense." As if to hedge his bets, Ward concluded that Rosie was truly "'designated for death' because her capacity to live her life is fatally compromised" by her poor cardiopulmonary system.[74] She had always been intended to die, while Gracie had not. George Annas noted in his review of the court's decision that Ward, before concluding, "condemned the parents' refusal to choose life for [Gracie] in dramatic terms: 'In my judgement, parents who are placed on the horns of such a terrible dilemma simply have to choose the lesser of their inevitable loss. . . . Parents with equal love for their twins would elect to save the stronger.'"[75]

The second appeals judge, Lord Justice Brooke, agreed with Ward that although the separation was in the best interests of Gracie, it could not be said to be in the best interests of Rosie. Rosie, he decided from the medical evidence, was a "reasonable creature" in the eyes of the law—that is, a living person capable of being murdered, as opposed to a subhuman parasite. Rather like Ward, he concluded that Rosie was, "sadly, self-designated for a very early death." And he agreed with Ward that the act of separation would be a "positive act," meaning that it would actively cause the death of Rosie, and was therefore an act in need of legal justification. But unlike Ward, Brooke found the legal justification for the sacrifice surgery not in the doctrine of self-defense but in the doctrine of "necessity": Rosie might not be attempting to kill Gracie, but she was clearly standing in the way of Gracie's survival, and so Rosie's death might be deemed necessary if Gracie's survival depended on it. Ward noted that the necessity defense could be used if all of the following requirements were satisfied:

1. the act [in this case, sacrifice surgery] is needed to avoid inevitable and irreparable evil [early death for Gracie];
2. no more should be done than is reasonably necessary for the purpose to be achieved [saving Gracie];

3. the evil inflicted [death to Rosie] must not be disproportionate to the evil avoided [death to both].[76]

Brooke decided "that all three of these requirements are satisfied in this case."[77] He therefore dismissed the parents' appeal.

The third appeals judge, Lord Justice Robert Walker, supported Judge Johnson's view that the separation might be in Rosie's interests: "To prolong [Rosie's] life for a few months would confer no benefit on her but would be to her disadvantage." Like Ward, Walker found justification for the sacrifice in "the doctors' duty to protect and save [Gracie's] life." But he insisted as well on "each twin's . . . right to physical integrity." "Each twin's right to life," Walker wrote, "includes the right to physical integrity—that is, the right to a whole body over which the individual will, on reaching an age of understanding, have autonomy and the right to self-determination." According to Walker, neither Rosie nor Gracie could have that physical integrity if left conjoined; and thus, to prevent the surgery was to deny their rights to privacy, autonomy, and self-determination. Offering a sort of consolation prize to Rosie, Walker declared: "The operation would give her, even in death, bodily integrity as a human being."[78] She would be killed, but in being separated she would at least get to become one of us.

Indeed, though the pattern was largely ignored in press accounts and subsequent analyses, the singleton assumption—that a life conjoined is an unjust, unworthy life—ran through the opinions of all three judges. Like Walker, Ward thought that Rosie had "a full claim to the dignity of independence," a dignity that could be given her only through separation. No matter if the process of acquiring dignity killed her; she must have it. Gracie had the same right, in Ward's view, regardless of how egregiously the surgery disabled her: "Whatever her residual disabilities, . . . they are likely to be slight in weight in comparison with the strength of her right to claim, as a human being, the dignity of her own free, separate body."[79] Brooke likewise tipped his hat to the singleton assumption: "The doctrine of the sanctity of life respects the integrity of the human body. The proposed operation would give these children's bodies the integrity which nature denied them."[80] Everyone ought to be a singleton, regardless of the cost; every sane person would want that.

In his review of the legal judgments for the *New England Journal of Medicine,* George Annas notes that "the case seems to have been decided not on the basis of the law (which most of the judges found of little help) but on an intuitive judgment that the state of being a conjoined twin is a disease and that separation is the indicated treatment, at least if such treatment affords one of the twins a chance to live."[81] Curiously, "three of the four judges believed that [Rosie] was better off dead than continuing to live for a few months as a conjoined twin."[82] Annas finds particularly problematic Ward's imaginary monologue in which Gracie declares, "Stop it, Rosie, you're killing me." Who knows what Gracie would really have thought if left to grow up conjoined? "Each twin might, . . . of course, consider the other twin to be an integral part of herself," writes Annas. And there is good reason to think this would have happened, given what conjoined people have said about their siblings. In any case, Annas concludes, "made-up monologues cannot take the place of legal analysis."[83]

Annas notes another important thread running through the appeals judges' decisions—namely, "a strong desire to authorize physicians to do what they think is best for their newborn patients."[84] All three judges seemed genuinely baffled as to why the parents were much more pessimistic about Gracie's prognosis than the doctors. The judges seemed to identify with the desires of the surgeons to intervene and produce a "normal" life for a "normal" child cursed by a fluke of nature. Walker voiced his leanings when he declared that "*highly skilled and conscientious doctors* believe that the best course, in the interests of both twins, is to undertake elective surgery in order to separate them and save" Gracie.[85] Ward revealed his primarily allegiances when he said that it would have been acceptable had the physicians decided with the parents against separation. Whatever the *physicians* wanted was all right with him. Annas writes: "The conclusion of Lord Justice Ward that it would have been 'perfectly acceptable' for physicians to decide either way must be wrong: if [Rosie] is a pursuer who is killing [Gracie], saving

[Gracie's] life (and that of others in her situation) by ending [Rosie's] life must be mandatory. The court's ruling that physicians can do what-ever they think is best (with the court's prior approval) is no legal rule at all."[86] One can infer from the three opinions that "the judges identi-fied strongly with the physicians and had little empathy with the par-ents or their religious beliefs."[87] The legal battle became, in the eyes of the winners, a tussle between modern, rational, professional judgment and premodern religiosity and sentimentality.

A number of ethicists joined Annas in declaring the court's decision wrongheaded. Most argued that it was ill advised not because sacrifice surgeries are morally flawed, but because overriding the parents' wishes in this case was inappropriate. Like many of his colleagues, Raanon Gillon, professor of medical ethics at Imperial College, London, argued that "there are good reasons for removing parental consent—when the parents are being negligent or when they have really weird views that would result in the deaths of their children. But these parents do not have really weird views. They have very standard views, the most impor-tant of which is you don't kill one person in order to save another." Yet Gillon and several of his colleagues stated that if they were in the place of Rina and Michaelangelo Attard, they would choose to pursue the op-eration. Gillon found "both ways . . . legitimate" and therefore believed that "both ways should be left to the parents to decide."[88] A number of physicians, including surgeons, publicly agreed with this conclusion. Dr. Keith Roberts, who had participated in separation surgeries himself, re-marked: "Neither the medical staff nor the lawyers are going to be faced with the problem of looking after what could be a severely handicapped child and a child that is going to be faced with numerous operations in the medium term."[89]

While "the court's decision did not mandate the surgery, . . . it held that the procedure would be lawful, should the doctors decide to pro-ceed."[90] Rina and Michaelangelo Attard could have pressed on, appeal-ing to the House of Lords or even to the European Court of Human Rights, but they declined to do so. After their capitulation, the director of the Pro-Life Alliance, Bruno Quintavalle, tried to wrest control of

Rosie's legal representation away from her court-appointed representative, but his request was denied.[91]

And so, on November 6–7, 2000, surgeons at St. Mary's Hospital performed the separation over the course of twenty hours. The act of cutting Rosie off from her blood supply "was carried out in respectful silence." Surgeons Alan Dickson and Adrian Bianchi made the fatal cuts together, feeling that "it was inappropriate for one person to shoulder the burden of consigning [Rosie] to death." Dr. Dickson, a Catholic, "said he prayed in his car on the way to operate on the twins." Dr. Bianchi, an Evangelical Christian, figured Rosie ended up in heaven, where she could watch over her sister.[92] "I feel that the 'ghost' of [Rosie] remained with us, as though she is helping her twin sister along."[93] The fact that these openly religious men were the ones to carry out the sacrifice did not mollify pro-life groups, which roundly condemned the procedure.

According to St. Mary's Hospital, after the twins had been separated a team of physicians and nurses tried, without success, to revive Rosie.[94] But they could not seriously have hoped to revive her; the very reason the operation had proceeded was that her cardiopulmonary system was inadequate. Rosie was buried two months later, on January 19, 2001, in Malta, after an inquest concluded that her death had followed upon "surgery separating her from her conjoined twin and that the surgery was permitted by an order of the High Court, confirmed by the Court of Appeal."[95] Her parents went to Malta briefly for the funeral, and then returned to England, where Gracie was still in the hospital. They recalled feeling that Rosie had been "freed by death." A month after the surgery, Michaelangelo told a television reporter: "Sometimes you just can't believe that she is dead, but you have to accept that she is. . . . It was good to hold her [after separation] because it was the first time we could cuddle her because she was always joined." His wife felt the same: "I wanted to see [Rosie] straightaway" after the operation. "I lifted and cuddled her. She was dead, but I was happy that I was holding her."[96]

In the operating room, as soon as Gracie had been "freed" from Rosie, the surgical team started the work on reconstructing Gracie.

Bianchi later recalled: "That involved bringing the pelvic bones to-gether, which, in effect, brought the legs into the right position, so now you had someone who looked like a baby."[97] It's not clear whether, after the surgery, the medical team felt the need to do anything for Gracie psychologically. The usual assumption is that at such an early age, a baby is not very aware of being conjoined or being separated. In an in-terview with the BBC News, Lewis Spitz, surgeon at London's Great Or-mond Street Hospital, said that at his institution, after separation "we put a mirror next to [the separated twins], giving them an image of something. That really helps them settle after surgery." Betraying his frustration that the separation had been carried out at a much less ex-perienced institution than his own, Spitz added: "That's the benefit of experience—that you have tricks like this."[98]

At the age of ten months, Gracie returned to Malta with her parents. Going home with their smiling daughter, Rina and Michaelangelo con-fessed that "in the end we are happy that the decision to separate was taken by the judges." They had come to realize that "of course we're now happy that we still have Gracie."[99] Like many parents, they seem to have underestimated their ability to cope with a disabled child. Reports about Gracie's health and prognosis vary widely, as is typical with such complex separations. Within a few days after the separation, "the baby rapidly improved" in terms of cardiovascular function, "as predicted by the surgical team."[100] The stress on her heart and lungs was greatly di-minished after the surgery. A month after the operation, Dr. Dickson told reporters he was "pleased with [Gracie's] progress, particularly as she 'looks so well and is so happy and so normal.'"[101] In June 2001, when she went home to Malta, Dr. Bianchi told the BBC that Gracie was using "her legs just like any other child. She is actually sitting up on her own, and she sits in her baby walker and pushes herself around."[102] The doctors at St. Mary's think that "eventually she could be of normal in-telligence, able to walk, have children and have an average life expec-tancy."[103]

Nevertheless, because she was joined so intimately and complexly to Rosie at her abdomen and genitals, and because her legs formed at such an acute angle to her spine, Gracie has needed and will continue to need "substantial surgery to reconstruct her lower abdomen, rectum, and . . . her sexual organs."[104] The malformation of her legs will require at least two surgeries to straighten them; "her lower spine must be reconstructed; her lower internal organs, rectum, and vagina need major reconstruction; and her bladder will have to be rebuilt."[105] She will be subject to the same kinds of genital reconstructive surgeries that children with intersex conditions have been through. The doctors assured the appeals court judges that such surgeries work very well, though their plans focused exclusively on providing Gracie with a "functional" (i.e., open) vagina so she could have intercourse, ignoring entirely the question of what these surgeries might do to her sexual sensation, a topic that never arose in the male-dominated discussions.

After the gag ordered was lifted by the judges, Rina and Michaelangelo sold their story and pictures to a news organization. The money went into a trust fund to pay for Gracie's surgeries and rehabilitative therapy. Rina and Michaelangelo Attard have assured reporters: "She's going to be a real fighter."[106]

One of the initial concerns expressed about the rulings in the Attard case was that the precedent might create a slippery slope: doctors might increasingly "kill" some people to "save" others. But the judges were careful to limit the scope of their decision, insisting that the ruling applied only "to cases where X cannot be saved without killing Y, where Y, by continuing to exist, would inevitably kill X, and where X is capable of an independent life but Y is not, whatever the medical intervention."[107] While I appreciate the judges' attempt to avoid setting an undesirable precedent, such a limitation marks conjoined twins as fundamentally different from all other people. It codifies the singleton assumption (a life conjoined is no life), which I think is wrong. I am not sure what decision I would have made in the Attard case as a parent, though I feel sure that as a physician or a judge I would have respected the parents' relationships with and sense of duty to both their daughters. The asymmetry between the girls was so much more marked than in the case of

Amy and Angela Lakeberg, that in some readings of Rosie's anatomy she does seem to drift from the side of "autosite" to the side of "parasite." But then the dismal reports about Rosie's brain function always came from those in favor of the sacrifice surgery, and I suspect that as a parent I would have had the same reaction the Attards did: I would have seen her as a true second daughter. This is, after all, how most parents feel about such a child, even if they know she will soon die.

I would very much like to see the medical evidence and legal judgment in the Attard case replayed in a somewhat different light—that is, with a consciousness of what conjoined twins have said about their own lives. And without the singleton assumption that being separated is necessarily better than being conjoined. Without the belief that compassion requires intervention and normalization. Without presumed consent where one cannot easily or legitimately presume it. In this altered light, the judges and doctors might have had a much harder time assuming that sacrifice was ethically and legally justifiable, whether or not the parents wanted it. They would have had a harder time creating the "X, Y, Z" exception that marked conjoined twins as a special case of personhood.

Let's consider a final case, one that barely made the news. No background of intense national debates about healthcare costs, no legal battle between parents and doctors. Again, the parents are Roman Catholic—not a coincidence, since Catholics are more likely to bring their conjoined twins to term. (In each of the three cases discussed in this chapter, the Catholic parents declined to consider abortion.)

Sandra and Ramon Soto, a couple from Puerto Rico, were in their twenties when they found themselves faced with the decision whether to sacrifice one child to save another. In June 1999 in Boston, Sandra gave birth to conjoined girls. During her pregnancy, doctors at Brigham and Women's Hospital and Children's Hospital of Boston had used the latest imaging techniques to determine that the twins were conjoined in such a way that, shortly after the umbilical cord was cut, both would die. The girls were joined at the chest. Darielis Milagro ("Miracle") Soto had a functioning, healthy heart on her side. But Sandra Ivellise Soto had no heart and no aorta, a condition known as acardia.[108] She was being kept alive by her sister's cardiovascular system through a rare anomaly that sometimes shows up in nonconjoined twins: twin reversed-arterial-perfusion sequence, known as a TRAP sequence.[109] Darielis Milagro was pumping blood to her acardic sister "via the umbilical cord. That meant that cutting the cord, normally a happy event in the delivery room, would kill both babies."[110] Sandra Ivellise would be cut off from the blood supply keeping her alive, and so she would die. If not separated immediately from her sister, Darielis Milagro would die too, probably within minutes—a few hours at the most.

The physicians made sure the expectant parents understood what they were proposing: "Dr. [Steven] Fishman gathered [the parents] and other family members at the hospital and drew a detailed diagram of the surgery on the blackboard. He finished with a large X over the outline of the twin who would die. 'We cried,' Mr. Soto said. 'I guess until then we thought God was going to put a little heart in there.'"[111] (The doctors could not hope to connect a transplant heart from an unrelated brain-dead donor, since Sandra Ivellise also lacked essential blood vessels.) Advised of the prospective sacrifice surgery, the parents consented, and the medical team prepared for the separation. On May 30, 1999, fairly late in the pregnancy, Sandra Soto's blood pressure rose to dangerous levels. A cesarean section was performed. "Fishman stood by in the delivery room at Brigham and Women's Hospital, waiting to rush the babies through the corridors to Children's Hospital, which is next door. But, he said later, he was unprepared for the emotional impact of seeing two live babies and knowing that soon there would be only one. 'As much as we spent months planning for this, and it seemed ethically and emotionally simple, nevertheless when I saw them both initially pink and both crying and moving their arms and having the same size bodies, it was heart-wrenching."[112]

Only a few minutes after delivery, Sandra Ivellise started to show signs of metabolic distress. Her skin was cooler than her twin's; "she had no pulse, and her blood pressure could not be measured. Immediate surgical separation was initiated."[113] Knowing that Sandra Ivellise would not survive the operation, the surgeons made the cut farther to her side, and used some of her ribs and skin to close Darielis Milagro's wound. Sandra Ivellise was buried in Puerto Rico, in a family plot. Darielis Milagro had two follow-up surgeries to address internal problems and was sent home with her parents at the age of six months.[114] "She also needed physical therapy and a brace to help straighten her spine, which was curved backward from her cramped position in the womb."[115] In August 2000, reporter Denise Grady, who broke the story, wrote: "In the end the gamble worked. Today the surviving baby girl is a healthy fourteen-month-old, with huge brown eyes and an impish grin, living with her parents in New Jersey."[116] Grady reported that after follow-up surgery to fix the chest bulge created by the original reconstruction, Darielis Milagro "looks completely normal, with surprisingly faint scarring and a remarkably sunny disposition, considering all she has been through." The total cost for her care has so far come to "more than half a million dollars, with portions paid by Medicaid programs in Massachusetts and New Jersey, and the rest absorbed by the hospitals." When Grady asked the Sotos what they would tell Darielis Milagro when she was older, they answered: "The truth."[117]

The natural reaction of a person with a typical anatomy is to look at a child like Gracie Attard or Darielis Milagro Soto and feel that the ends have justified the means. Any criticism of the means must surely be mere selfish philosophy in the face of these smiling children. At least, I find myself feeling this way. Yet what keeps bothering me is knowing, as a historian, that the apparent successes of Gracie and Darielis will have a disproportionate influence on decisions about separation surgeries, including sacrifice surgeries. They will influence these decisions much more strongly than our knowledge of the lives and deaths of Amy Lakeberg, Angela Lakeberg, Rosie Attard, and Sandra Ivellise Soto— will influence them much more strongly than our knowledge of what conjoined twins have said about the way they wish to live and die. In the history of pediatric surgeries, the history of those who have been counted as successes has always weighed more heavily than the history of those who have been counted as failures, including those who have remained conjoined. And every day, the pace of technological interventions in pediatric medicine picks up because in practice even one small "success" makes any doubt—ethical or empirical, personal or social— feel like a sacrifice of hope.

Freeing the Irish Giant

4

The movie *Liebe Perla* (1999) documents an old woman's quest to locate film footage depicting her and her family. In the 1930s, in Hungary, Perla Ovitz and several members of her family formed a traveling musical act. They weren't exactly like the Von Trapps: in addition to sharing a love of music and performance, they also shared a type of dwarfism. But the footage Perla Ovitz is trying to locate in *Liebe Perla* did not focus on her family's act. It was shot by Dr. Josef Mengele's staff at the Auschwitz death camp for the purposes of scientific study. Ovitz vividly remembers standing with her family for a very long time, all of them naked, while the camera rolled.

Mengele's obsession with shaping a perfect race drove his interest in "little people"; he thought that in studying them he might crack the code for human stature and find a way to engineer taller bodies. In one of the most disturbing moments of *Liebe Perla*, Ovitz remembers Mengele with a sort gratitude. "I can't say anything bad about him," she confesses, pointing out that Mengele's interest in her family's anatomy led him to become, in a way, their protector. "But for him, we too would have died."[1] It is true that without Mengele and his fascination with dwarfism, Perla Ovitz would probably have been sent to the gas chambers. Not only was she "defective" in terms of her stature, but she was Jewish, a "race" the Nazis thought of as anatomically flawed.[2]

*   *   *

The museum of the Royal College of Surgeons in London—known as the Hunterian Museum after its founder, the anatomist John Hunter (1728–1793)—is of modest size but contains a priceless collection: preserved human and animal remains, some of them excellent specimens of common anatomical states, some of them holding an important place in the history of medical progress, some of them examples of very unusual anatomies. For instance, there is the double skull of the Two-Headed Boy of Bengal, the only specimen of its kind: a case of parasitic conjoinment in which an individual had, conjoined to the top of his head, another incompletely developed head (see Figure 9).[3]

Near the museum's entry hall is the skeleton of Charles Byrne, who

9.
The Two-Headed Boy of Bengal, who was born in 1783 with an incomplete conjoined twin attached to his head.

went by the stage name "O'Brien, the Irish Giant." When he died in 1783, at the age of twenty-two, he was almost eight feet tall. Public appearances enabled Byrne to earn a great deal of money, travel widely, and meet interesting and famous people, among them doctors and anatomists interested in his condition. Byrne had been born in Ireland; his father and mother (both of unremarkable height) were Irish and Scottish, respectively. But why had he changed his name from Byrne to O'Brien and added the adjective "Irish" when exhibiting himself in England? He was using a practice followed by many other anatomically unusual performers of the eighteenth and nineteenth centuries (including Chang and Eng Bunker, who billed themselves as the Siamese Twins). Adopting foreign-sounding names increased their exoticism and, probably more important, reassured spectators that the anomalies before them were the traits of an "inferior" race, not something that happened to the best people.

Eyewitness accounts and advertisements indicate that Byrne engaged in a relatively genteel form of exhibition, appearing in "elegant" rooms for a limited number of hours each day.[4] He charged various amounts, depending on the ebb and flow of public interest and on the spectators' ability to pay. A newspaper advertisement from August 12, 1782, contains this notice:

Just arrived in London, and to be seen in an elegant apartment, at the cane-shop, in Spring Garden–gate, . . . the Living Colossus, or wonderful Irish Giant, only twenty-one years of age, measures eight feet two inches high. [Byrne's height was often exaggerated for promotional purposes.] This extraordinary young man has been seen by [an] abundance of the nobility and gentry, likewise of the faculty, Royal Society, and other admirers of natural curiosities, who allow [i.e., claim] him to surpass anything of the same kind ever afforded to the public. His address is singular and pleasing, his person truly shaped and proportioned to his height, and affords an agreeable surprise. . . . Ladies and gentlemen are respectfully informed that the hours of admittance are from eleven in the morning till four in the afternoon and from six to seven in the evening every day, Sundays excepted. Admittance 2s. 6d.[5]

Though Byrne made a small fortune through such engagements, he was apparently not a particularly sophisticated man where money was concerned. In 1783 he turned his savings of seven hundred and seventy pounds into two large bank notes, one for seven hundred pounds and the other for the rest, and soon thereafter found himself easily robbed of his estate. He died not much later, perhaps because of drinking too much, a habit "to which he was always addicted."[6]

The Hunterian Museum's records reveal how Byrne's skeleton came into the possession of John Hunter. Byrne was aware that anatomists were eager to acquire and dissect his body when he died—a fate he very much wanted to avoid. Perhaps he felt it would put him in a league with criminals, whose corpses were commonly used for dissections. Perhaps, as a Christian, he worried that on Judgment Day his body would not be resurrected intact if it had been dissected. Perhaps he simply feared the scalpel. Whatever the case, he left a will stipulating that his money be used to pay for a burial at sea, where his body would be safe from science.

But when Byrne died, a bidding war broke out among the anatomists. The undertakers, who were about to dispose of the corpse, sold it to the highest bidder—namely, John Hunter. Hunter dissected the body, wrote a scientific report on his findings, and prepared Byrne's skeleton for display. It is now a prized artifact in the Hunterian Museum's collection.

By starting this chapter with these two stories, I don't mean to imply that the Royal College of Surgeons is akin to the Nazi Josef Mengele. Mengele captured, tortured, and killed his human subjects, seeking to build a master race for purposes of world domination. The Royal College of Surgeons oversees the training and practice of surgeons, seeking to advance healthcare so that patients may live better lives. About as enormous a difference in practice and mission as you can find.

What I do mean to suggest is that the history of the display of people with unusual anatomies has not been an especially a pretty one.[7] And this matters, because people's social and political identities and their sense of self are constrained to a large degree by how they—and people like them—are represented by others. If, for example, you are fat, and if virtually all of the representations of fat people in the mass media and the medical literature equate people who are fat with laziness, pathology, and failure of will, you are apt to be seen by others and to see yourself as lazy, pathological, and weak-willed. If you are an adult who is very tall (a giant) or very short (a dwarf), and if most of the representations of such people portray them as freakish, diseased, or, at best, fantastic, you are apt to be seen by others and see yourself as freakish, diseased, or fantastic. And the perceptions of parents and doctors who make decisions about genetic screening, selective abortion, and normalization treatments will be influenced by the same context.

By juxtaposing the stories of Perla Ovitz and Charles Byrne, I also mean to show that the involvement of scientific and medical professionals in the examining or displaying of people with unusual anatomies unfortunately does not guarantee conformance with the noblest values of biomedicine. Professionals may not always have the patients' or subjects' best interests in mind. Indeed, because biomedical professionals are often concerned with the prevention and normalization of unusual anatomies, their relationship to people with such anatomies may contain an unresolved tension, perhaps even an irresolvable conflict.[8] A person with an unusual anatomy often sees her anomaly as an integral and valued aspect of who she is, even if it causes physical or social pain for herself or others. How, then, can the biomedical professional accept that individual and her complex claims about her self, while at the same time using her to figure out how to remove or prevent future instances of that valued aspect? It's a tough position for both parties.

You could give the response I've encountered on the professional side: "These are deformities we're talking about! Only if you go mad with political correctness can you be troubled by the idea of treating them as such." But that argument collapses if you reflect on why Josef Mengele brought Jews as well as dwarfs to Auschwitz—namely, to get rid of anatomical "defectives"—and if you realize that Mengele and his colleagues, when seeking ideas about how to deal with "defectives," looked to America, land of the free and home of the brave.[9] In the 1920s and '30s, the United States led the world in scientifically advanced ideas on reducing the incidence of inborn mental and physical "deformities" in the general population.[10] Many Americans, liberals as well as conservatives, saw eugenics—including forced sterilizations and euthanasia—as both just and scientific. In other words, the approach taken by the Nazis was extreme, but an extreme manifestation of a scientific trend sweeping the Western world.

The definition of "birth defect" turns out to be anything but simple, if you think historically. Medical journals and books published before the twentieth century contain images of "birth defects" very different from the ones seen today. Recently, a woman with an intersex condition made a telling comment on such modern-day images of people like her: they look, she said, like "insects tacked to a board for study."[11] By contrast, pictures in pre-twentieth-century biomedical texts look more like family portraits done at Sears. These drawings, engravings, and (after about 1830) photos of conjoined twins, giants, dwarfs, hermaphrodites, and other people with atypical anatomies usually show the individuals standing in a stately looking room, their faces unobscured, their eyes often meeting the viewer's gaze (see, for example, Figure 10). Unless there was some specific reason for portraying them nude—say, the researcher wanted to show the breasts and penis on a hermaphrodite—the subjects are clothed. If they *are* nude, they're often in classical poses. Certainly these pictures show the influence of artistic standards of the time. But they do more than that. They also show that biomedical professionals saw their subjects in a certain way, subtly different from the view that today's professionals have of their subjects, who are typically depicted nude, in stark clinical settings, eyes and faces masked out or blurred. And those old images convey a subtly different message about the subjects.

How were the relationships between biomedical professionals and people with unusual anatomies different in the nineteenth century? A careful reading of the advertisements for Charles Byrne reveals that, while making a living from exhibition, "O'Brien the Giant" associated with medical and scientific men for his own financial gain. "This ex-

10. Charles Byrne, the Irish Giant, shown with two other giants and a number of people with dwarfism in a late eighteenth-century etching. The spectators include Lord Monboddo, William Richardson, Mr. Bell, and Bailie Kyd.

traordinary young man has been seen by [an] abundance of the nobility and gentry, likewise o*f the faculty, Royal Society, and other admirers of natural curiosities.*"[12] Indeed, from the late eighteenth to the mid-nineteenth century, people with atypical anatomies frequently participated in a tacit exchange of goods and services: they let curious medical and scientific experts examine them, and the professionals not only enjoyed some free voyeurism (voyeurism has always been part of the attraction), but they also published accounts of those anatomies, increasing medical knowledge of the conditions and enhancing their own reputations. In exchange, the professionals gave performers like Byrne expert opinions about their conditions, occasional medical treatment, and spoken and written testimonies to their strangeness—opinions that were as important and prestigious as those of the "nobility and gentry," with whom men of science were commonly grouped. Performers could then use these medical testimonies in advertisements and penny pamphlets to drum up more business.

This notion of fair exchange comes through clearly in comments by Dr. William H. Pancoast, who attended the Philadelphia autopsy of the Bunker twins. As Pancoast remarked in his report to the College of Physicians in Philadelphia: "To advance their own interests [Chang and Eng] frequently consulted medical men in different parts of America and Europe, as to the safety of a surgical operation to divide the band and release them from their peculiar connection; [but] these consultations [with medical men] were mainly used to excite the curiosity of the public, as it is believed by those who knew them well, that they never, except once, seriously contemplated such an operation." Pancoast considered it "a duty to science and humanity" that—in exchange for the doctors' sustained help in "exciting the curiosity of the public"—the family of Chang and Eng should permit an autopsy. For "the twins had availed themselves most freely of the services of our profession in both hemispheres, and it was considered by many but as a proper and necessary return" that the medical men should at last be allowed to satisfy their curiosity about the Bunkers' internal anatomy.[13] Biographies of Millie and Christina McCoy, who were contemporaries of the Bunkers, also show the way in which the opportunity to examine an atypical anatomy was traded for salable expert testimony. At an early age, the Two-Headed Nightingale "Millie-Christina" was brought to New Orleans for a command performance,

> in obedience to a request from the medical faculty of that city, [who asked] that she be brought there for scientific examination. Rooms were taken and every preparation made for the contemplated examination, after which she was to be placed on public exhibition. . . . The examination . . . at length took place and proved most satisfactory, every physician in attendance concurring in pronouncing her Nature's greatest wonder. *Being endorsed by the medical faculty,* she was now put on public exhibition.[14]

The sisters repeatedly and without charge performed their songs and dances privately for doctors, and, in exchange, were given valuable endorsements (see Figure 11). Their penny pamphlets included "certificates of eminent medical men," a distinction they boasted of in their theme song:

> Two heads, four arms, four feet,
> All in one perfect body meet,
> I am most wonderfully made
> All scientific men have said.[15]

Relying on the respectability conferred by such testimony, performers were able to spice up their penny pamphlets with information on their sexual anatomy, details which would otherwise have been considered lewd. Presented in the form of a straightforward quotation from two medical doctors, a description of Millie and Christina's sexual anatomy could be included in their pamphlets. It was safe and acceptable for the public to read that Millie and Christina had "separate bladders, but one common vagina, one uterus to be recognized, and one perfect anus,"[16] so long as this information came from a medical professional. In this way, doctors could gentrify and legitimate a performance that might

11. Millie and Christina McCoy, the Two-Headed Nightingale, as shown on the cover of a pamphlet sold in conjunction with their exhibition tour in 1869.

otherwise be simply distasteful. Medical discourse was deliberately used to ward off charges of pornography, even while it was used to titillate.

Robert Bogdan has noted that many so-called freaks of the nineteenth century earned enough money by exhibiting themselves to lead financially secure lives.[17] The trade in medical testimonies obviously helped. Today, by contrast, although unusual anatomies are often displayed in medical schools and textbooks, on television, on the Web, and in the popular press, the profits do not accrue chiefly or directly to the exhibited. And "respectable" people find most nonmedicalized displays of unusual anatomies pathetic, exploitative, or distasteful.

Now, I am not looking to suggest that we return to the freak show era, or to suggest that doctors "help" their patients by providing them with testimonies of how odd they look. We do not know whether most nineteenth-century performers would have chosen this means of profit if an alternative had been available, or whether many would choose it today if it were a real option. I am hardly romantic about the great age of exhibitions. But I do find it remarkable that there was a time when doctors were quite *publicly* thrilled to obtain an audience with these patients; a time when they would, in penny pamphlets as well as in the medical literature, celebrate them as extraordinary, bizarre, amazing; a time when they would recognize these people as authorities on a unique and strangely attractive experience.

But a critical shift occurred in the latter part of the nineteenth century, as medicine became more prestigious and more aligned with science. Physicians formerly had been quite willing to exchange concrete and enthusiastic testimonies for access to particular interesting bodies and the personal accounts that came with those bodies. Now physicians began to offer a much more abstract value—"the good of humanity"—in exchange for ready and unlimited access to *all* unusual bodies. Over the course of the nineteenth century, we see a fading of the idea that the biomedical professional should have to give something immediate in exchange for access to interesting anatomies. Today, such professionals tend to feel a primary *right* to see and use and own unusual anatomies—whether these be skeletal remains, extraordinary genes, or patients deemed (in insider lingo) "fascinomas." Professionals claim this right of access not because they have given the unusual person something equally valuable in return, but because, in a very abstract and universalized sense, science and medicine supposedly serve all of humanity.

From time to time, we hear news stories about anatomical specimens (usually from races other than white, and cultures other than Western) that are being repatriated from museums to their ancestral lands. The moral of these stories is pretty clear: science is being robbed of its rightful access to the rare and unusual anatomy, and progress in the field is being impeded. Part of me always does feel indignant on behalf of the scientists. The link between modern representatives of the culture and the people whose bones are being transferred can seem extremely tenuous, and the loss to science great.[18] But then the education I've received from people with unusual anatomies always forces me to think of the other side. How does the shift in ownership change the balance of power? Who ought to own that body and the stories which can be drawn from it? (Authority emerges, after all, from an "author.") Why must all believers in scientific progress necessarily agree that the mission of such progress is more important than the representation of peoples with whom the specimens share important links?

One of the many "services to humanity" that biomedical science seeks to perform is the prevention and normalization of unusual anatomies. This is the very reason biomedicine often gets free and easy access to those anatomies. But think of the irony. People with unusual anatomies hear medical professionals saying: "We get to see you, examine you, and display you at will, because we're trying hard to fix you and to prevent anyone else from being born like you." Of course, what they are really trying to prevent and alleviate is *suffering.* But when one's identity is grounded in the experience of one's anatomy—as it is for virtually all of us—the elimination of that experience can legitimately be equated with the elimination of the self.

As a historian of medicine, I sometimes feel that people with unusual anatomies have been freed from the realm of circus sideshows only to be caught in a circus-like realm of medicine—all in the name of "service to humanity." Just as medicine once gentrified and legitimated Millie and Christina's pamphlets, it now seems to have become the acceptable—even required—venue for the display of deformity. Everybody is entitled to get in on the show, so long as the show involves surgical sutures and latex gloves. (One specialist in twin separations has stated that a frequent logistical problem with such operations is the "large number of personnel who want to view [the] surgery.")[19] It becomes increasingly difficult for people with unusual anatomies to live without having their lives constantly fitted by others into the story of medical normalization. Have they tried medical normalization? If not, are they planning to do so? If they don't want to be normalized, why on earth not? Even "radical" documentaries about people with unusual anatomies—films that take place entirely out of the medical setting—always bring up these questions, and in so doing medicalize the individuals' differences. Wouldn't it be odd if every biopic of a normate examined the subject's medical and cosmetic history?

In a critical analysis of *Katie and Eilish,* a film about the early life and separation of the Holton girls, Catherine Myser and David L. Clark reveal how documentaries about medical normalizations echo the freak shows of bygone eras. Such shows, they argue, "not only claimed to disseminate a certain folk wisdom about human and animal wonders, but also reiterated normative expectations about the boundaries dividing the titillatingly 'freakish' and the reassuringly 'ordinary.'" In this way, they resemble films about medical normalizations.[20] By focusing on how a "deformed" child is to be made "normal"—how conjoined twins are made into singletons, for example—medical documentaries reinforce the idea that the unusual anatomical state is unjustly imprisoning the *real* child. By implication, the real child always has a typical body; at best, a child with unusual anatomy is seen as an unfinished product that requires someone else's expertise to become fully human.

These stories become much more troublesome when they focus on children from developing countries who are brought to North America or Europe for high-tech normalization. Anyone conversant with the history of freak shows—who knows that being of a "primitive" race was often enough to land a person in a Western exhibition—will find these tales uncomfortably familiar. Yet such films avoid disturbing the general public by cloaking themselves in the mythology of a quest narrative: the brave and heroic child undergoes great trials to achieve a noble destiny.[21] Indeed, the narrators and adult figures in these films commonly attribute the normalization decision to the child, though the child is always too young to have expressed a choice. In the film *Katie and Eilish,* after Katie dies as a result of the operation, her father protests: "We feel that Katie *wanted* us to take that chance. . . . She *wanted* to be separated."[22]

Documentaries about pediatric normalizations do feel very different from old-style freak shows, in that they are permeated not only with charity and devotion but also with the hope of redemption from "unjust" embodiment. Perhaps it is the lack of redemptive tone that repels a lot of viewers when talk shows feature adults with unusual anatomies. The guests on these talk shows are typically open and positive about their experiences, often vigorously objecting when a host tries to re-inject a redemptive tone by lauding them as exceptionally courageous "super crips."[23] These days, people frequently alert me when conjoined twins are going to be featured on some talk show. I used to avoid watching those shows. I would cringe (the usual middle-class intellectual response) and feel sorry for the guests. But then I began asking people with unusual anatomies how they felt when they appeared on the shows, and nearly all of them found it an extremely positive, even empowering experience. Having been taught to hide their "shameful" anatomical difference, they saw such shows as an opportunity to be "out" and proud of who they are. They were treated by host and audience as authorities on an important experience, as people worth talking and listening to—as *respectable adults.* And just as in the old-fashioned freak shows, on the talk shows audience members frequently got a chance to speak directly with the attraction, sometimes asking questions so ignorant or obnoxious that the star had the chance to appear clearly superior to the dolt or jerk. While medical documentaries about normalizations tend to infantilize people with unusual anatomies, talk shows often do exactly the opposite, raising issues of sexuality, professional oc-cupation, and the like.

Modern-day talk shows are also a bit like the freak shows in that they sometimes pay the person with the unusual anatomy, sharing some of the profits. By contrast, when documentaries, prime-time news maga-zines, and popular-press segments feature people with unusual anato-mies, the subjects are almost never paid, despite the fact that the film-makers, journalists, doctors, medical institutions, and news agencies stand to profit financially or through a boost in reputation, which the person with the unusual anatomy doesn't. This unequal distribution of profits is perpetuated in the name of journalistic integrity.

Not incidentally, some people I spoke with explicitly contrasted ap-pearing on a talk show with being displayed in a teaching hospital. In the latter case, they felt exposed (and often were, since they were usually presented in a state of undress), silenced, pathologized, and often pitied, without a chance to respond. Though the medical professionals did not intend this, the subjects often felt that they were asked to function pri-marily as specimens of tragic biological mistakes, of the medical pro-fession's charity, of a surgeon's triumphant handiwork, or all of the above.[24]

One of the greatest ironies in the history of the doctor-patient rela-tionship can be seen in the trend toward making patients anonymous in professional publications. There was a time when patients were identi-fied in the medical literature with at least their first names or initials, and sometimes, in the case of unique anatomies, their whole names. Sometimes the reader was told where the person had been born, lived, worked, and died. If a picture was included, the person's face was often clearly visible. Today, all of these practices would be considered unethi-cal, a breach of doctor-patient confidentiality. But the irony is that the newer system, in attempting to respect the privacy of the patient, has turned the focus away from the individual to the condition itself, sup-planting the life and person with the "deformity." This might be better for medical science, of course, since science seeks objectivity, but it is surely problematic for medical practice.

The custom of masking patients' eyes and names continues, even when the idea of maintaining patient confidentiality is patently absurd. The medical report on the separation of Katie and Eilish Holton published in the *Journal of Pediatric Surgery* refers to them as "Twin 1" and "Twin 2." In the nude photo of them, their eyes are masked with black rectangles.[25] But anyone who knows anything about conjoined twinning knows that the report is about Katie and Eilish; they had been shown with their real names in news reports all over the world, and the detailed description of their bodies and histories are unmistakable. Intersex advocate Cheryl Chase is absolutely right when she says that the black rectangle over the eyes accomplishes only one thing: it saves the viewer from having to endure the gaze of the subject.[26] The viewer is always able to recognize the subject if the subject is already familiar. (I've met many people who, while researching their own conditions, discovered what they immediately recognized as their own "objectified" pictures and stories in medical texts.)[27] Having consented to many news features about the family, the Holtons might well have consented to let their daughters' names and unmasked photos be published in the *Journal of Pediatric Surgery*. Instead, by forcing Katie and Eilish into a hyperobjectifying pose, linguistically ("Twin 1," "Twin 2") and photographically (nude with eyes masked), they were protecting the *doctors* from scrutiny, not the twins. Such visual and verbal language, which deprives the patients of personality and authority, forestalls or silences questions about what happened to them and why.[28] In this sense, while a masked photo can be termed customary, it cannot really be termed respectful.

Of course, there are clear signs of progress within medical educational and clinical settings. More and more medical schools are inviting people with unusual anatomies, not to be physically examined as fascinomas but to speak as experts, imparting valuable information on disability and difference. The subject who returns the viewer's gaze is making a comeback. Instead of including page after page of dehumanizing and depressing pictures and stories of people with unusual anatomies, some genetics and pediatrics textbooks now contain "real-life" photos and stories supplied by the subjects. Imagine how differently people (medical students, expectant parents, and so on) would view conjoinment if medical books contained not only clinical photos but family-life scenes, such as the photo of the Hensels reproduced here as Figure 5. Is such an image irrelevant to medical practice and medical science? I don't think so. After all, conjoinment is understood to be primarily a psychosocial problem. Why, then, provide a decontextualized nude photo as medical evidence rather than photos of conjoined people in some of their daily social situations? The latter seems more accurate in some ways than the former, though the former is painted with a veneer of objectivity.

Especially powerful are new books in which a person with an unusual anatomy provides an expert critical analysis of the medical-treatment system. This, more than anything, puts the author in a position of authority alongside those with the power to offer and execute preventions and cures.[29] Also progressive (and subversive) is the increasing visibility of medical professionals with unusual anatomies—a visibility aided by the 1990 Americans with Disabilities Act, which has helped to remove some of the institutional obstacles barring the way of disabled people who wish to become medical professionals. Lisa Abelow Hedley's documentary *Dwarfs: Not a Fairytale* includes a portrayal of Michael Ain, a pediatric orthopedic surgeon at Johns Hopkins University who has achondroplasia.[30] The popular television show *ER* featured a similar character: an attending physician who always used a crutch. Such individuals, who combine the voice of biomedical authority with an unusual or disabled anatomy, help to ensure that their colleagues will not be (in Adrienne Asch's words) "distracted by disability"—that is, apt "to ascribe negative aspects of a disabled person's life solely to the biological characteristics of the condition."[31] They might then understand why "an increasing number of people with disabilities are coming to believe that their problems reside largely in society rather than in their atypical biology . . . [and] that clinicians' attitudes toward disability often perpetuate negative stereotypes and exacerbate the 'difference' of disability."[32]

Another hopeful sign: some healthcare providers and medical associations are supporting autobiographical works by artists with unusual anatomies. For example, the American Academy of Orthopaedic Surgeons has actively promoted the drawings of Laura Ferguson, an artist with scoliosis. Ferguson's self-portraits are exquisitely subversive—executed in a style similar to that found in many eighteenth-century anatomy texts, while explicitly authored by the patient herself and containing a clear element of eroticism (see Figure 12). People who see Ferguson's work are drawn to it but also provoked by it. Her autobiographical "Visible Skeleton" series poses a radical question: Can a scoliotic skeleton be physically painful and gorgeous at the same time? A challenging question like this has the power to help patients with scoliosis and other conditions begin to reject the social stigma assigned their bodies—begin to sort out various kinds of pain and the options for addressing them.[33]

A new breed of documentary is likewise indicating progress. These films follow the day-to-day lives of people with unusual anatomies, yet dispense with the medicalized rhetoric. A superb example is Ellen Weissbrod's film *Face to Face: The Story of the Schappell Twins*,[34] which has aired many times on the Arts and Entertainment channel. It portrays the thirty-seven-year-old Lori and Reba at home, at play, at work, and on visit to New York City, while avoiding "the clichéd trope of triumph over adversity—all too common in narratives of disability."[35] The film does not give the impression that the twins are brave heroes or freaks of nature. If anything, it makes their lives seem remarkably unremarkable. After observing them in their ordinary activities, one realizes that they're typical in every way but the obvious. As G. Thomas Couser notes, *Face to Face* treats Lori and Reba's conjoinment "less as a physical impairment than as a condition that constitutes their lives and selves as tightly, permanently, and complexly intertwined. Their constant, prolonged exposure itself has a normalizing effect." By the end of the film, it is obvious that they are sisters "who seem to get along supremely well and to enjoy each other, who do not pity themselves and who do not regard themselves as freaks or in any way unattractive."[36]

Instead of distancing the viewer from the twins, Weissbrod draws

12. *Crouching Figure with Visible Skeleton,* by Laura Ferguson. A self-portrait in oils, bronze powder, charcoal, colored pencil, pastel, and oil crayon on paper, 12.75 in. × 10 in., dated 2000.

parallels in subtle but highly effective ways. Lori and Reba get up together, get dressed, eat breakfast; all viewers "who have shared bathrooms with siblings, spouses, partners, parents or children will recognize that the differences in privacy are matters of degree, not of kind."[37] Lori and Reba declare that they are not broken and don't need fixing. In the same context with this statement, people on the street are asked to talk about what they would change about their *own* bodies. All the interviewees name anatomical features that cause them social stress—though none of them say they have plans to alter those features. In one sequence, a couple who have obviously been married a long time insist that conjoinment is nothing like marriage, because marriage is something that is voluntary; and then they "sheepishly realize that they are finishing each other's sentences," sounding as if they were one being.[38]

In contrast to most medical documentaries, *Face to Face* keeps the talking heads to a minimum, using them mainly to provide the narrative structure needed to connect its points. I am one of those talking heads, though I was initially reluctant to participate, because I was tired of assisting on various medical docu-dramas. (Even if you're posed as a critic in such films, as I usually am, you're still helping to produce them.) But it quickly became clear to me that Ellen Weissbrod really *was* going to do something different, as different as A&E could handle, something much more in keeping with the way conjoined twins themselves have spoken about their lives. I knew the film was going to be *really* different when she asked me if I would dance with other "characters" in the film; the footage would appear in the music video she was making for the end of *Face to Face*—a sequence in which Reba, an aspiring country music singer, performs Reba MacEntire's "Fear of Being Alone." At first I thought Weissbrod was nuts—and not because she knew what a bad dancer I am. But then I realized that her idea was brilliant: the film, which seeks to dissolve the glass wall separating viewer and subject, would conclude by dissolving the line between the typically disempowered subject and the typically empowered medical expert. And so I joined the film crew and participants in a park in Pennsylvania and did my best to dance.

One of the sharpest edges in Weissbrod's film is its treatment of staring and gawking. Most documentaries about people with unusual anatomies contain footage of strangers gawking, but the footage is left essentially unexamined. It provides only a sad undercurrent, or sometimes a challenge to the improbable claims of the subject that she or he is doing fine, thank you very much. But *Face to Face* confronts the issue of staring head-on and stares it down. On the trip to New York which begins the film, Weissbrod gives Lori and then Reba a video camera like the film crew's, so that we can see what the twins are seeing and thus get a vantage point from which to examine our own perspectives. When particularly boorish folks seek to pat Lori and Reba like good-luck charms, children, or pets, or take pictures of them out of "scientific" interest, Lori and Reba vocally challenge them, asserting their right to be taken seriously as individual *people*. Reba (Couser writes), "the more outspoken of the two, asserts that any revulsion at the sight of them in public is not their problem but the spectator's. As she puts it, and [Lori] concurs enthusiastically, if you find the twins interesting, you are welcome to use binoculars; if you don't, you should use blinkers."[39]

Neither Lori nor Reba (nor, for that matter, Weissbrod) implies that the simple act of looking at them is pathological or mean. All seem to acknowledge that it would be really surprising if people *didn't* look. A journalist once spent some time with Lori and Reba, and then asked me: "Don't you think it's terrible how people look and stare?" I told her it's perfectly natural to want to take a good, long look at someone whose anatomy is like nothing you've ever seen. Surely our brains are set up to categorize; this feature is useful for survival and makes life a lot easier. So why wouldn't we do a mental cartwheel when we see someone who seems to be more than some*one*? My two-year-old son once came upon my research notebooks, which included pictures of conjoined people, and even he wanted to know as much as possible about what he was seeing. The problem comes when we look away as if there is some terrible shame—which there shouldn't be—or treat people as if they exist *merely* to satisfy our curiosity. To force oneself to look away is, I believe, to act like those who institutionalized Lori and Reba for the first twenty-four years of their lives: it is to assert that people like Lori and Reba are not fit to be seen in the world.

There's obviously something incredibly attractive in certain unusual anatomies. If there weren't, exhibitions and freak shows wouldn't have survived as long as they did, talk shows wouldn't keep running specials on these subjects, A&E wouldn't have funded *Face to Face,* and the public wouldn't keep going to the Hunterian Museum to see the skeleton of Charles Byrne and to the Mütter Museum to see the plaster cast of the Bunker twins. For me, the problem lies not in the attraction but in the shame ascribed to the attraction—a shame that is nowhere evident when someone is drawn to a person with a conventionally gorgeous anatomy.[40] The shame always shuts out the possibility of recognizing and analyzing the feelings of attraction, or of using that attraction to build respect of anatomical difference.

So when people automatically condemn efforts by those with unusual anatomies to make money from others' attraction, I think they are wrong. Of course there are forms of exhibition and performance which are degrading to all parties (dwarf-tossing comes readily to mind). But there are also forms of performance which undercut moronic stereotypes and oppressive assumptions. And if they offer no money to the performer, they end up looking a little too much like charity balls: displays of pity, instead of confrontations with the real issues.

Danny Black makes his living as an actor, an entertainer, and a distributor of products for people with dwarfism and other forms of short stature. He is an outgoing, witty, gregarious man with a healthy sense of irony. One of the T-shirts he sells reads "Midget Petting Zoo," simultaneously mocking the word "midget" and the obnoxious habit some people have of "petting" short adults, as if they were children or good-luck charms. On the topic of disability rights, Danny is extremely knowledgeable and holds strong views. And this doesn't make sense to a lot of people. How can he be aware of the way society discriminates against people like him, yet also make his living off people's fascination with his differentness? For one thing, he loves being a professional performer—he's exceptionally good at it, always making people feel at ease—and he sees no reason why he shouldn't "use something that I've been given and use it to its fullest extent."[41] The work helps him pay the bills (though it is not enough to cover the cost of health insurance); and despite the provisions of the Americans with Disabilities Act, Danny has found that discrimination makes other forms of employment hard. But he also loves performing because his work gives him an opportunity to engage people socially and politically.

When he attends a fraternity or bachelor party as a paid guest, Danny invariably finds himself talking with people who otherwise would be unlikely to meet someone with an unusual anatomy. The first question is always: "What's the right term for you people? I've heard that one should say 'little people,' and not 'dwarf' or 'midget.'" Danny then moves these new acquaintances from a discussion of labeling to a personal conversation, and in the process they end up losing—or at least loosening—their assumptions that dwarfism is a tragedy, a mythical state, or a joke. Yet Danny doesn't see his work as simple missionary work. He truly loves performing, and appreciates that it allows him to be fully "out" about who he is. His car is hard to miss, emblazoned as it is with the words "ShortDwarf.com," the name of his company. Indeed, I think it is Danny's forthrightness about being a dwarf that upsets some people, who wish that he wouldn't (as he puts it) "use my disadvantage to my advantage." Some people seem to want him to feel at least a bit ashamed about who he is and what he does, to spend more time in the closet, to hide behind politically palatable terms, and to work on "passing" in normate society when passing is possible.[42] He isn't interested in doing that. For Danny, making money off of other's curiosity allows him to establish a certain authority, to stop being a mere subject, to open a dialogue—and to make a living when many other employment paths continue to be closed because of social prejudice. Seeing him perform is rather like watching a dwarf skeleton in a stodgy old anatomy museum suddenly grow flesh, emit a laugh, and walk out of the glass cage.

In 1998, when I was in London for a conference on body-enhancing technologies and human identity, I took my colleague and friend Tod Chambers to the Hunterian Museum. For a historian of anatomy the

Hunterian is Mecca, and I thought that Tod, who teaches ethics at Northwestern University's medical school, would love it as much as I do. I now remember with embarrassment how I enthusiastically showed him various famous remains and excellent specimens, and approached the case displaying Charles Byrne's skeleton with the same unexamined enthusiasm. But Tod paused to read the museum's account of how it had acquired the skeleton, and he asked me why I wasn't deeply disturbed by the situation. Why wasn't I horrified, as he was, that the Royal College of Surgeons was displaying the skeleton in utter disregard of Byrne's wishes? He began quoting my own work back to me, reminding me of my objections to the way people with unusual anatomies had historically been treated by the medical profession, and wondering why I seemed to be in "gee whiz" mode. He pointed out that the display had little to do with the size and might of Byrne and everything to do with the size and the might of surgeons.

I suppose it had never occurred to me to question the display of Byrne's skeleton because I saw it—the skeleton, the display, the story—as an institution. But I had Tod's critique in my head a few days later, at the University of Cambridge, where I gave a presentation on my work. I included some brief comments on the skeleton, and soon found that I was being followed around by a rather unusual woman who kept encouraging me to champion the "rights of the dead" and get Byrne's remains buried. The situation felt rather like a *Monty Python* skit, in part because her accent added a comedic touch to the refrain "The rights of the dead!" Comedy aside, this didn't seem the right way to think about the situation. I didn't feel that my work was about the dead.

Back home, I thought about writing to the Hunterian Museum but I didn't do so immediately. Again, the issue of the skeleton didn't seem to relate to my work. Moreover, I had a selfish fear of annoying curators, who had the power to restrict access to the collections, and whom I liked and counted among my friends. But Tod kept sending me notes encouraging me to "free the Irish Giant!" So in June 1999 I finally wrote to the Keeper of the collections and asked why the museum wasn't following Byrne's wishes. I heard nothing back, and when I sent a follow-up letter I was told vaguely that the curators were taking it under advisement.

In 2000 A&E aired Ellen Weissbrod's *Face to Face,* and one of the people who contacted me afterward was J. Bruce Beckwith. A pediatric pathologist, he was the co-discoverer of Beckwith-Weidemann Syndrome (BWS), a growth anomaly, and was the researcher who defined and named Sudden Infant Death Syndrome (SIDS). Throughout his professional life he had helped families in their efforts to deal with pediatric illnesses and unusual anatomies. He had started the first parental support group for SIDS in 1965, and in retirement he continued to work with a BWS support group. In view of his personal and professional qualities, I was honored that he thought well of my comments in *Face to Face.* He particularly appreciated my view that "surgeons are often too quick to separate twins that might better remain together, out of the bias that only separateness can be good, no matter what the cost in lost anatomy and physiology that surgery would entail."[43] Maybe, I thought, with his long history of enlightened practice, Bruce would help me to get Byrne buried.

So in May 2000 I asked him what he thought of the situation at the Hunterian. I described my ongoing discussion with Tod Chambers and said that, as "a lover and protector of anatomy museums," I did not wholeheartedly want to see Byrne buried. But the more Tod and I talked about it, "the more I agreed that this was a case in which a man's explicit wishes about his body were being ignored."[44] Bruce completely disagreed:

The central issue here is this: What are the rights of the deceased, and how do they compare to those of the living? As I pondered this issue, I reflected upon the fact that in my career as a pathologist, including many years as a part-time forensic pathologist, I have often been obliged to violate the wishes, both express and implied, of the deceased concerning the treatment and disposition of their mortal remains. A patient may have had strong religious, cultural, or personal objections to the performance of an autopsy, but had the misfortune to die under circumstances where the legal system, and/or the interests of the living public, mandated that an autopsy be performed. On hundreds of occasions I have been the perpetrator of such autopsies.[45]

Pondering Bruce's words, I finally figured out why this way of thinking about the situation—as an issue of the "rights of the dead"—felt wrong to me. I wrote back:

> *I* see the case of Byrne as representative of a larger issue—what the living are and aren't able to control about their own bodies in the medical and scientific arenas. It seems to me especially common to ignore or discount the wishes of those with unusual anatomies (be they teratological or pathological [in origin]). I feel as if what has happened to Byrne is just of a piece with what happens to people who are living with other unusual anatomies—intersex, for example.[46]

So when Bruce encouraged me to give up thoughts of freeing the Irish Giant—to "stick to the living"—I felt this was exactly what I *was* doing. I was working for the living. Byrne was dead, yes, but all of the people taking harmful lessons from that display were quite obviously alive.

Keeping Charles Byrne's skeleton on display sends a terrible message about the modern-day relationship between physicians (especially surgeons) and people with unusual anatomies. It says that such people do not have an equal say in their fate, that they can be readily exhibited to the public as symbols of nature's freakery or medicine's miracles, whether or not they wish to be. Moreover, it says that being considered a freak is enough to exempt a person from the social norm of respect, especially in dealings with medical and scientific professionals. So in trying to get Byrne's remains buried, I was partly concerned for the well-being of people living with unusual anatomies. But I was also concerned for the well-being of medical science and medical practice. If we want the public to honor and provide support for medical science and medical practice, biomedical institutions need to avoid cases of blatant exploitation. If museums of anatomy are to be museums of medical progress, then they need to signal not only an understanding of how medical diagnosis and treatment have advanced, but also an understanding of what constitutes progress in doctor-patient relationships and medical ethics.

But Bruce did not agree with me on this point. And in some ways I could relate to his gut reaction when he wrote:

> Twice have I stood before . . . Byrne's skeletal remains, and both times shivers of emotion have gone through me. . . . He represents, in a way, the tens of thousands of corpses disinterred by physicians and medical students of earlier generations in the interests of advancing science and instructing young physicians. Tens of thousands of sad, revolting stories. But for most of the victims of those crimes, the good that came from the study of their bodies represents their most lasting contribution to human welfare. I would vote to let [Byrne's] bones rest where they are. His spirit departed from the bones more than two centuries ago, but his bones still have the ability to make good people shiver, and cause us to reflect upon the tragedies that have accompanied so many human advances.[47]

Bruce's instinct was to defend the institution of medicine and biomedical science. And, I realized, I was now at the point where my instinct was to defend people born with unusual anatomies. Suddenly I realized that the defenders of biomedicine and the defenders of people with unusual anatomies had little chance of coming together. This seemed bizarre to me, since doctors who treat children with "deformities" are classically seen as paragons of beneficence. I wasn't ready to give up the dialogue, and responded:

> We have to consider the special case of Byrne, the fact that he was a person with an unusual anatomy, and recognize that it is possible that—if real, legitimate scientific studies of his skeleton were [finally] to be carried out at this point—the aim of information garnered from his bones would probably be to prevent or eliminate the very condition that made him unique, famous, wealthy, and an object of Hunter's desires. And would Byrne want his bones to be used to prevent or eliminate giantism? Possibly, but very possibly not. My point here is not that we should ro-

manticize giantism or similar conditions. [Giantism can be associated with physical and psychosocial pain and suffering.] Rather, we have to recognize that Byrne's condition was not a horrible one, and therefore this is a case where there are concerns neither of public safety nor justice that should override Byrne's desire to be buried where the anatomists could not reach him.[48]

After writing another follow-up inquiry, I received a slightly more substantial reply from the Hunterian Museum. The Keeper of the collections wanted to know on whose behalf I was making the request. I understood the question: usually requests for repatriation (which is what this request looked like) come from compatriots or family. I was neither. My response: "I am asking on behalf of people concerned that physicians follow the wishes of individuals before and after their deaths."[49] It seemed to me that the failure of John Hunter in 1783 to follow Charles Byrne's wishes was one thing, but it was quite another for the Royal College of Surgeons, at the turn of the millennium, to persist in this course.

The Royal College remained unmoved. And for my part, I've pretty much given up trying to bury the remains of Charles Byrne—my efforts have irritated too many people I respect. But I haven't given up on the essential understanding that Tod and Bruce (and Laura Ferguson and Cheryl Chase and Danny Black) helped me to acquire, and so I retain the notion of "freeing the Irish Giant" as a metaphor. Not as a metaphor for emptying out anatomical museums; the vast majority of their exhibits do not derive from people who tried, as Byrne did, to use their life's earnings to avoid ending up in glass cases. Rather, I think of "freeing the Irish Giant" as a metaphor for ending particularly damaging and wholly unnecessary medicalized displays of unusual anatomies, displays which harm the image not only of people with unusual anatomies but also of doctors and scientists genuinely interested in reducing suffering. After all, the culture at large takes its cues about how to feel and act about today's Charles Byrnes and Danny Blacks and Katie Holtons not only from skeletons in museum cases, but from the people who put and keep them there. So I'm going to keep trying to persuade doctors and scientists—who are so right in wanting to manifest beneficence—to free their Irish Giants. And on days when I need a good laugh, I think of what Winston Churchill used to say to himself: History will vindicate me, and I will write the history.

The Future of Anatomy

5

People often ask me: Do we have good reason to believe that conjoined twins are being detected in the womb and aborted ever more frequently, so that there will be fewer people born conjoined in the future? I think the answer to that specific question must be yes,[1] but if the question is intended (as it often seems) to be a larger inquiry about the future of the experience of conjoinment, I think it fails to recognize the breadth and depth of the social context of conjoinment. The experience of being conjoined *will* be different in the future from what it was in the past, but not because there may be fewer people born conjoined; conjoinment is so infrequent that the numbers scarcely matter either way. The future experience of being conjoined will be different because many realities of anatomy are changing—and they are changing for all of us.

Consider, for example, how often today we are *all* encouraged to think of the socially problematic aspects of our various identities as simple matters of health and medicine. Lab-coated women at Clinique (read: clinic) counters all across the land push scientific-sounding cosmetics designed to shrink pores and even out skin tone, constantly assuring customers this is about the health of one's skin. Drug companies recommend that you "ask your doctor" about Rogaine, Claritin, and Viagra, seeming to offer a medical choice, while the images in such direct-to-consumer advertisements make clear what is really at stake—

namely, establishing a favorable identity (handsome and virile man, attentive wife and mother, limber and sexy senior citizen). And although many of us try to stay thin because obsessing about the bathroom scale pushes us up the social scale, corporate advertisers and the press encourage us to pretend that fat is simply a disease, or at least a disease waiting to happen; "treatments" for weight loss are invariably wrapped in a veneer of medical technology, even when they may be more dangerous than healthful.[2] In the United States, the values of individualism, self-improvement, free enterprise, and high-tech medicine have combined in the past few decades to create a culture in which one is able—indeed, even expected—to employ medical technologies to alter one's anatomy and make it more socially advantageous.[3] The same trend can be seen in reproductive technologies, where genetic screening, prenatal testing (including blood tests and sonograms), selective abortion, and pre-implantation embryo selection are now routinely employed by prospective parents, who are given technological and social encouragement to avoid having children who might present socially challenging anatomies.[4]

Now, this scenario—in which questions about social identity become medicalized, and surgeons, geneticists, and other medical professionals mediate the relationships between anatomies and identities, even the most superficial—has been long in coming. For at least a hundred years, people in the West have turned to doctors to help them figure out, in cases of ambiguity, who is a man and who is a woman, whether a body contains one person or two, which kind of body represents a life worth saving. Even before the days of generally safe "corrective" cosmetic surgeries, doctors were given the cultural authority to decide who would have which identity based on anatomy, and those that didn't fit cultural norms were labeled pathological or defective.[5] Surely at some level all medical professionals, whether they mean to or not, help to construct social anatomical norms, as well as people's personal and social identities. A person often will suffer social stigma just from being sick; so retaining or regaining physical health doesn't just feel better—it may improve an individual's social standing.

But if questions about the degree to which medical professions *should* be adjudicating issues of social identity are not new, they certainly have greater urgency today, for it is getting harder and harder to draw a basic philosophical distinction between the clinic and Clinique. Though doctors and nurses have primary responsibility for aiding us when we are sick or dying, and though their training focuses on these transgressive anatomical states, their professions are increasingly responsible for making people *look* better. Of course, medical professionals are being given this role in part because breast implants, Botox injections, and the like carry significant risks with them; I'm not sure anyone (except maybe those who stand to profit) wants to see a system other than one in which these procedures are approved by the FDA and conducted only by trained medical professionals. But the effect of this system is to increasingly medicalize, and thereby legitimize, cosmetic procedures designed to make people's anatomies more socially valuable. And because health is such a strong social value, and having a stigmatized identity is increasingly equated with having an illness, more and more people—not just those born with fairly unusual anatomies—are being encouraged to seek "medical treatments" for what may otherwise be considered benign anatomical variations. The cycle of anatomical shame attribution and medical normalization seems to be accelerating for all of us.

The shift toward the medicalization of less-than-ideal anatomies may be particularly nefarious because of the naive yet pervasive belief that the shift is just a matter of health and individual choice. You find out about the risks and benefits and you decide what's right for you or your child, right? Isn't it all about autonomy, the right to choose? Hardly. One's options are clearly delimited by what is offered, how it is offered, how it is viewed, what is affordable, and what others are choosing. With anatomical normalizations (minor and major) increasingly available, increasingly suggested, and increasingly allied with the prestige of health and modern medicine, it will become more difficult and more costly for people to resist them. So the makers of Botox Cosmetic disingenuously assure us that "it's really up to you. You can choose to live with wrinkles.

Or you can choose to live without them." Although technically the choice to use Botox injections may be up to us, the fact that (some) people now have a choice has radically changed our world. It was not an option for anyone before; and the more it is expected, the less of a free choice it becomes.

Now, I don't want to suggest that *all* of this "normalizing" and "enhancing" medicine is socially regressive. At first blush, it may seem as if the politics of appearance should never be medicalized—that it is dangerous to have doctors participating in the high-stakes social game of who counts as beautiful or normal. But the medicalization of some important aspect of individual identity can be a relatively positive thing for an unjustly oppressed person. Ronald Bayer has shown how gay men and lesbians were moved into the clinic before they moved themselves beyond that, to public-pride events.[6] The mid-twentieth-century medical treatment of homosexuality as a disease was a visible improvement over the treatment of homosexuality as a heinous crime. Although some of the medical treatments of homosexuality—including the transplantation of "healthy" testes (those of heterosexual men) into gay men, clitorectomies of lesbian women, electroshock therapy, and various brutal psychotherapies—were horrifying, they were perhaps less horrifying than the treatment of gay men and lesbian women by the justice system. Most important, the medicalization of homosexuality began to raise questions about the supposed choice of homosexual identity; and by questioning the nature of the "disorder," it asked whether there really was anything wrong with being gay or lesbian.

Similarly, many women have found that the medicalization of "premenstrual syndrome" has enabled them to deal with it more effectively; the medicalization of PMS (or, lately, PMD) has validated their cyclic physical and emotional tension, and thereby increased their options for coping with it.[7] As we saw in Chapter 2, some parents have been relieved to see their children's unusual anatomical variations medicalized. The fact that many people with personal experience of alcoholism support the search for an "alcoholism gene," and the fact that many transgendered people hope that researchers will find a unique variation in their brains, are evidence that the medicalization of a stigmatized or shamed identity can feel (and even *be*) personally and socially liberating.

But this kind of liberation through biomedicine doesn't work for everyone. I mean this literally: there is unequal access to medicine, especially medicine deemed cosmetic. I also mean it figuratively. The liberation doesn't work for people with unusual anatomies, because the medicalization of identity often leads to the medicalization of their entire existence—their lives are increasingly categorized in terms of their potential for medical normalization. They are marked as diseased, as broken, where one might otherwise see a diseased and broken social system. In a world where the anatomical basis for identity is medicalized and striking variations from the ideal are often pathologized, the first question asked about a newborn is not the general one—What can we do to give this child a good life?—but a much narrower one: What can we do to make this child normal? The consequence for many people born with unusual anatomies is that, paradoxically, their unusual anatomy—no matter how much it becomes part of their identities—is perpetually viewed by others as a transitory state, an inferior stage of development.

If you look around a bit (particularly in cities, where most social movements start) you'll see some corrections in the trend toward medicalized anatomical ideals. There's a counterculture in the United States that values unusual anatomies—cyborg enhancements, intentional transgenderism (the use of hormones and surgeries to produce blends of male and female parts), body piercing and body art, blatantly artificial hair coloring and styling. But since these practices are favored by a marginalized set of people, and since, however permanent these body modifications are, they are clearly *chosen* by the subject, I doubt that this adoption of unusual anatomies will do much to improve the social and medical treatment of people born with unusual anatomies.

I put more faith for positive change in the disability rights movement, which has done much to counteract the medicalization of social normality. Now, I say this in spite of the fact that families and individuals coping with unusual anatomies have often remarked to me that an unusual anatomy does not in itself constitute a disability. For example, when I asked Lori Schappell if she is disabled, she answered, "No, . . . I can do anything. I can climb a ladder."[8] For years I thought the same thing—that having unusual-looking genitalia or a cleft lip didn't disable a person, because it didn't involve any impairment of physical function considered species-typical. But after years of studying the social and medical treatment of unusual anatomies, I now believe it makes sense to think of many unusual anatomies as disabilities, even when they involve no physical impairment in the customary sense. Many unusual anatomies are treated almost exactly like traditional disabilities. That is, a physical difference is treated as a physical defect; "pity-ridden paternalism"[9] and medical "cures" are substituted for genuine accommodation; the subject's body instead of the body-environment interaction is treated as the fundamental problem; intense, debilitating, silencing shame is attributed where none ought to be; and basic rights—to employment, self-determination, education, marriage—are denied by others, who assume that these rights are the preserve of people with typical bodies. Kathi Wolfe could easily be describing people who are conjoined when she writes—with reference to people with disabilities—that they are caught in a "matrix of sentiment, stereotype, ignorance, and curiosity. . . . People see us as beggars, helpless victims, or superheroes."[10] In his classic book on the origins of the disabilities rights movement, Joseph P. Shapiro describes the understanding the movement is trying to promote: that the problem is not the individual body but the need "to fight someone else's reality."[11] The description could apply equally well to the intersex rights movement.

The disability rights movement can be an effective advocate for people with unusual anatomies because it is indeed a movement about *rights*. Like the women's rights movement, the civil rights movement, and the gay and lesbian rights movement, it challenges social and institutional restrictions placed on people because of their anatomical differences. Like these other movements, it strives to make the public aware that supposed natural barriers to education, mating, voting, and so on are really socially constructed barriers which can be altered. Why should a person be treated as if she were broken or abnormal or tragic just because she has a socially challenging anatomy? People without disabilities (sometimes termed the "not-yet disabled") often look at those with disabilities and think that disability is solely a function of the individual body: "she can't get around because she's blind" or "he can't go to school because he can't use his legs." But physical abilities are determined by the interaction of our bodies and our environments, and this is true for all of us. Society simply accommodates some bodies better than others.[12]

Moreover, as disability rights activists have pointed out, the same sorts of claims about "individual impairment" and "natural defects" were used to justify limiting the rights of women and blacks. Nineteenth-century biomedical experts claimed that the "naturally inferior" anatomy and physiology of women and blacks relegated them to inferior social status. In 1851 Samuel Cartwright reminded medical professionals that "anatomy and physiology have been interrogated, and the response is, that the Ethiopian, or Canaanite [i.e., the African], is unfitted, from his organization and the physiological laws predicated on that organization, for the responsible duties of a free man." The person who was of African descent was, "like the child, . . . only fitted for a state of dependence and subordination. . . . The great primary truth, *that the negro is a slave by nature,* and can never be happy, industrious, moral or religious, in any other condition than that one he was intended to fill, is of great importance to the theologian, the statesman, and to all those who are at heart seeking to promote his temporal and future welfare." It was the African's inherent impairment, not the social situation, that held him back.[13] Cartwright's views were shared by many who believed blacks were not encompassed in the principle that "all men are created equal." Similarly, in 1850 a physician named N. Williams defended the idea that women should not become doctors. "The temperament of females," he wrote, "is less favorable for the medical profession than that of males. . . . There is a material difference in this respect. To the female sex, then, may be ascribed the *nervous* or *excitable* temperament. . . . The male sex is the most favorably constituted." Again, the argument focused on the natural impairment of the individual as a barrier to any attempts at social change: "Habit and education may do much to improve the temperament for this or any other department; but it cannot wholly supply the deficiency, or render the *artificial* arrangement of things equal to the *natural*."[14] These kinds of arguments about natural impairment—often couched as medical claims—were used to justify denying women the right to vote, to hold property, to make decisions about their own bodies.

Like lesbian women, gay men, interracial people, and exceptionally powerful women, people with unusual anatomies have been considered troublesome because they don't fit the rules about anatomy and identity. We have two genders (girl/woman and boy/man); therefore, people are supposed to come in two sexes (female and male) and mate heterosexually, the "natural" way. But gays, lesbians, and people with intersex conditions have been seen as problematic because they don't fit that simplistic model of sex and sexuality. Conjoined people have been seen as problematic because individuality typically equates with one person per body. People with dwarfism have been seen as problematic because adults are supposed to be taller than children. The modern-day solution has been to fix the apparent problem: the transgressive body.

But why go that way? Why not change minds instead of bodies? Why not—like the campaigners on behalf of women's rights and civil rights and gay and lesbian rights—think about the social nature of the problem of conjoinment (or intersex or dwarfism) and start there? Every day, science is unearthing more evidence that strict anatomical categories are human inventions, not natural phenomena.[15] That our thinking about such categories has evolved can be seen in the way the concept of race has altered. Scientists once thought that there were exactly five (or seven, or ten) human races, that each had its own characteristics, and that some were inherently superior to others, morally, intellectually, and physically. But it has been shown that there is more genetic and anatomical variation within any specific race than there is between that race and another. Labeling someone according to race tells you very little about that particular individual.[16] And thus there are now laws to protect people from discrimination on the basis of historically degraded racial groupings.

Similarly, although the most widespread notion is that there are only two sexes, and although there obviously is one common cluster of anatomical variations we categorize as male and another we categorize as female, nature doesn't tell us where to draw the lines. Nature doesn't decide how small a penis has to be before a newborn counts as intersexed instead of male, and nature doesn't decide that testes and a Y-chromosome make an individual a male even though androgen insensitivity makes the person look more classically womanly than most women. Search all you want for some particular gene, some particular chromosome, some particular hormone or brain-cell cluster—the fact is, *people* decide who will be grouped in what sex category, because it is important to do so for social reasons. Labeling someone according to gender will not allow you to predict that individual's abilities or potential. And thus there are now laws to protect people from discrimination on the basis of sex.

So why should people with unusual anatomies be treated as if their socially challenging bodies are inherently diseased? Why should the banner headline FREE AT LAST be used to announce the separation of conjoined infants?[17] People who are conjoined have repeatedly declared that bigotry about their conjoinment, not the conjoinment itself, is primarily what limits them. The clinic need not be the best or last stop for anyone with a socially problematic anatomy. Medicine could be sought to treat pain, metabolic dysfunction, and serious threats to health. But instead of seeking ambitious medical normalizations for children who cannot consent, we could seek social and legal liberation for people with unusual anatomies who are constricted because of oppressive assumptions about their worth and potential. Like the movements on behalf of women, people of color, gay men, and lesbians, a pride movement agitating for the benefit of people with unusual anatomies would have at its core a demand that people be treated not according to the color or the *shape* of their skin, but according to the content of their character. It would insist that the solution to being anatomically unusual is not to be treated with normalizations chosen by someone other than the subject, but to be guaranteed the same basic rights as others, including the right to determine for oneself what will happen to one's flesh when there is no medical emergency and when important physical capacities are at risk.

A protest march by people with unusual anatomies may strike some as a freak parade—but it is worth keeping in mind that the same slur was used against suffragists, civil rights protesters, and gay rights activists. Likewise, the medical and social treatment of children and adults with unusual anatomies may seem to be above criticism because it is beneficent in intention—but there was a time when people saw legal paternalism for women and enslavement for blacks as above criticism because those, too, could be beneficent in intention. In 1851 Samuel Cartwright commented on the "strange" tendency of slaves to try to escape: "With the advantages of proper medical advice, strictly followed, this troublesome practice that many negroes have of running away, can be almost entirely prevented. . . . They have only to be . . . treated like children, with care, kindness, attention and humanity, to prevent and cure them from running away."[18] Cartwright was not alone in this view. Doctors and scientists commonly favored the social oppression of women, black folks, and gay people, thinking this was the right thing for those individuals as well as for society.

The few brave souls who spoke up were deemed radical, unprofessional, or insane. Historically, the medical profession as a whole did not shift until progress had been made in the wider society. So if significant positive social change is to come for people with unusual anatomies, it will almost certainly have to start somewhere other than in hospitals and doctors' offices. There is no question that doctors, nurses, and other medical professionals are well-intentioned in their treatment of people with unusual anatomies; but they typically feel that the best they can do is adjust individual problematic anatomies to fit social expectations. They believe you must change children's bodies because you can't change others' minds.

But change *is* possible. I've seen this in attitudes toward intersex conditions. In the early 1990s, every major medical center and children's hospital in the United States was treating intersex the same way: with secrecy about diagnoses, with silences around doubts, and with surgeries and hormone treatments designed to make children with intersex look and act "normal." Back then, when campaigners for intersex rights questioned doctors, the response was: "The best thing we can do for these children is help them fit into society's expectations of sex and gender. In an ideal world we might do something different, but we don't live in an ideal world." Yet since the early 1990s, largely though the work of activists and journalists, tens of millions of people in the in the United States have heard and talked about intersex. In 1996, whenever I described my work to casual acquaintances, nine times out of ten I had to explain the most basic truths about intersex. Today, explanations are needed about one in ten times. Many expectant parents now know that intersex conditions are a possibility for their children. Many people now know that intersex is the mysterious "it" that occurred in their own (or their siblings' or cousins' or friends') lives. The context of the experience of intersex has changed radically for the better.

That said, medical treatment for intersex remains largely as it was in the early 1990s, when the intersex rights movement began. There is a little less secrecy about diagnosis, a little more hesitation about turning boys with "micropenis" into girls; but as I write, it is still mostly the same. Children with intersex still have their genital sensation, continence, fertility, health, and lives put at risk because of the belief that they will wither and die—socially and emotionally, and perhaps physically (by suicide)—if allowed to grow up with the genitals with which they were born. Specialists treating intersex still tell me, "You can't change society." But there are cracks in the system—cracks that, if enough people apply pressure, will become chasms and then canyons and then inhabited plains. When I started my work on intersex, adults with intersex who were openly unhappy with the care they received were not given a chance to speak to the experts. When I got a chance to do so on their behalf, I would suggest revisions in the standard of care: assign a gender as boy or girl soon after birth, recognizing that all such assignments are preliminary; hold off on unnecessary normalizing procedures; provide open, honest, shame-less psychosocial support for family and child. And I would be told that my suggestions were naively optimistic, even unrealistic. But nowadays adults with intersex conditions, along with other critics of prevailing practice, are taken seriously at major professional meetings. And at those conferences and in the medical literature, there are more and more physician leaders in intersex care unequivocally coming out against normalizing procedures that unnecessarily risk patients' health and functioning.[19] This is particularly true in the United Kingdom, where much greater value is placed on the need for strong evidence before risky intervention.[20]

There is also much positive leadership in the field of craniofacial care, particularly in the treatment of cleft lip and cleft palate. Whenever possible, craniofacial care providers adopt a team approach, recognizing that craniofacial anomalies (like many unusual anatomies) are chronic conditions that will not be "fixed" with a single surgery. Craniofacial teams now include not only surgeons but psychologists, social workers, audiologists, orthodontists, nurses, nutritionists, and dentists; they integrate ongoing peer support for patients and their families. Particularly impressive is the University of Washington Craniofacial Center at Children's Hospital in Seattle, where the team addresses patients' quality of life instead of simply assuming that a good cosmetic fix will mean a high quality of life and a poor cosmetic outcome will mean a low quality of life. The Seattle team includes Cassandra Aspinall, a social worker who was born with a cleft lip and who is the mother of a child with cleft; she has done much to institutionalize the powerful combination of personal and professional experience. Wendy E. Mouradian, a physician on the team, has urged that doctors shift their understanding of families "from deficits to strengths [in order] to promote health and well being in patients with congenital and acquired craniofacial conditions."[21] She emphasizes the need for long-term quality-of-life outcome studies, and recommends that professionals treating people with stigmatized craniofacial variations read (and share with patients) what past patients have written and said about their experiences. Rather than seeing patients' narratives as complaints or threats, Mouradian and some other physicians view these narratives as correctives to the pathologizing representations available in most medical texts. They are also mining patients' narratives for insight into how one can begin, in the words of Rosemarie Garland Thomson, "to critique the politics of appearance that governs our interpretation of physical difference, to suggest that disability requires accommodation rather than compensation, and to shift our conception of disability from pathology to identity."[22]

These changes enable us to envision a different future—for people with unusual anatomies, for their parents and children, for their doctors, for their neighbors, friends, and lovers. In this sense, a different future for us all, one in which technological intervention is no longer the primary means for demonstrating caring. Of course, many people with typical anatomies would have to help to fully realize this dream. Women didn't get the vote just because they wanted it; they got it because many of them wanted it and enough men made it possible. Likewise, civil rights legislation would never have passed without the help of whites in power. Maybe we're nearing the point where those of us born with more typical anatomies can truly rethink and reinvent how we enable others to become one of us.

To get to that point, we will have to come to a full understanding of the socially constructed nature of disability. Perhaps it is that symbol for "handicapped accessible" (really a symbol for "wheelchair accessible") which leads people to assume disability is only about a person's ability or inability to enter certain places. Too many people, including many of those who could legitimately be called disabled, think disability is about legs that don't move rather than the failure to build ramps.[23] In thinking that disability is about how far an individual can extend her leg—as opposed to the way an individual's anatomy is affected by the interaction of body and environment, including oppressive societal assumptions—people with unusual anatomies (and their families and doctors) often fail to see how the disability rights movement concerns them. People with unusual anatomies need, first and foremost, a pride movement that would encourage them to see themselves as full citizens and as a group with certain social problems in common. Raising their political consciousness would benefit not only them, but their families and doctors as well. And I would suggest that that consciousness-raising start where most do: with an understanding of the historical basis for the oppression. Once we recognize the historical basis, a different future becomes imaginable.

I recently looked up the definition of "individuality" and learned that there is an archaic meaning of the term: indivisibility. Archaic, indeed! Today singletons have a hard time imagining that being conjoined could ever function as an integral part of a person's individuality. At least one surgeon who does separations is accustomed to declaring in public, "Conjoined twins are born to be separated."

But what if we understood such twins as people who are no more broken than anyone else? What if we stopped thinking of biological anomalies as sworn enemies of humanity, and started recognizing their full social nature, perhaps even their social potential? In the long run, we can do better than try to guarantee every child a "normal" body.[24] We can try to guarantee a just world. If you take seriously what conjoined people have said about their bodies and their lives, you realize they are still experiencing what Mary Wollstonecraft felt in the late eighteenth century: "It is justice, not charity, that is wanting in the world." Let us now stop referring to children who undergo massive normalizations as "real fighters," and start recognizing that we are the ones who construct what they are fighting against.

Notes

Introduction

1. Lori Schappell, personal communication, December 9, 2002. The scene described took place in a liquor store when Reba, Lori's sister, tried to make a purchase.

2. Cheryl Chase, personal communication, November 25, 2002.

3. Danny Black, personal communication, December 13, 2002.

4. Ruta Sharangpani, personal communication, November 13, 2002.

5. The classic discussion of stigma and shame is Erving Goffman's *Stigma: Notes on the Management of Spoiled Identity* (New York: Touchstone, 1963). Goffman's sociological insights remain quite persuasive, though the book is more than forty years old and though his tone occasionally betrays a problematic contempt for his subjects. For an analysis of *Stigma* from a disability studies perspective, see Rosemarie Garland Thomson, *Extraordinary Bodies: Figuring Physical Disability in American Culture and Literature* (New York: Columbia University Press, 1997), pp. 30–32.

6. See Susan Wendell, *The Rejected Body: Feminist Philosophical Reflections on Disability* (New York: Routledge, 1996), ch. 4; see also Thomson's use of the work of anthropologist Mary Douglas (*Extraordinary Bodies*, pp. 33–38).

7. On the attribution of shame, see Goffman, *Stigma*, pp. 7–10.

8. Lori Schappell, personal communication, December 9, 2002.

9. Ruta Sharangpani, "Pity and Other Green Monsters," unpublished essay, quoted with permission.

10. Quoted in Kenneth Miller, "Together Forever," *Life*, April 1996, p. 56.

11. Janice Hopkins Tanne, "Free at Last," *New York Magazine*, November 15, 1993, pp. 54–62.

12. Keep in mind that, for reasons discussed in Chapter 2, most separations are done before patients are old enough to give their consent.

13. The term "normate" was coined by Rosemarie Garland Thomson to refer to people typically considered "normal." See *Extraordinary Bodies*, p. 8.

14. W. E. B. DuBois, *The Souls of Black Folk* (New York: Dover, 1994 [1903]), p. v.

15. See the website of the Intersex Society of North America (on whose board I serve voluntarily), www.isna.org. For Chase's account of the start of the intersex movement, see Cheryl Chase, "Hermaphrodites with Attitude: Mapping the Emergence of Intersex Political Activism," *GLQ: A Journal of Gay and Lesbian Studies*, 4 (1998): 189–211.

16. Cheryl Chase, "Affronting Reason," in Dawn Atkins, ed., *Looking Queer: Image and Identity in Lesbian, Bisexual, Gay and Transgendered Communities* (Binghamton, N.Y.: Haworth, 1998), pp. 205–219.

17. For a documented summary and critique of this treatment system, see Alice Domurat Dreger, "'Ambiguous Sex'—or Ambivalent Medicine? Ethical Issues in the Medical Treatment of Intersexuality," *Hastings Center Report*, 28, no. 3 (May–June 1998): 24–35.

18. See Alice Domurat Dreger, *Hermaphrodites and the Medical Invention of Sex* (Cambridge, Mass.: Harvard University Press, 1998). See also John Money, *Hermaphroditism: An Inquiry into the Nature of a Human Paradox* (Doctoral dissertation, Harvard University, 1952).

19. See "Special Issue on Intersexuality," ed. Cheryl Chase, *Chrysalis: The Journal of Transgressive Gender Identities*, (Fall–Winter 1997); Suzanne J. Kessler, *Lessons from the Intersexed* (New Brunswick, N.J.: Rutgers University Press, 1998); Alice Domurat Dreger, ed., *Intersex in the Age of Ethics* (Frederick, Md.: University Publishing Group, 1999).

20. See the thirty-minute video "Hermaphrodites Speak!" (San Francisco: Intersex Society of North America, 1997).

21. See 18 U.S.C. 116 (United States Code). See also Cheryl Chase, "'Cultural Practice' or 'Reconstructive Surgery'? U.S. Genital Cutting, the

Intersex Movement, and Media Double Standards," in Stanlie M. James and Claire C. Robertson, eds., *Genital Cutting and Transnational Sisterhood: Disputing U.S. Polemics* (Urbana: University of Illinois Press, 2002), pp. 126–151.

22. I first presented this story of the "Double X Syndrome" in Alice Dreger, "When Medicine Goes Too Far in the Pursuit of Normality," *New York Times*, July 28, 1998, p. B-10; reprinted in *Health Ethics Today*, 10 (August 1999): 2–5.

23. The Americans with Disabilities Act (ADA), Public Law 101–336, July 26, 1990, 104 Stat. 327.

24. On the relations among race, sex, and (dis)ability, see Adrienne Asch, "Critical Race Theory, Feminism, and Disability: Reflections on Social Justice and Personal Identity," *Ohio State Law Journal*, 62, no. 1 (2001): 391–423. For one man's account of how his life changed when he became black, see Gregory Howard Williams, *Life on the Color Line* (New York: Plume, 1995).

25. Comments by Kenneth Glassberg on *Dateline NBC* segment entitled "I Gotta Be Me," first aired June 17, 1997. Transcript produced by Burrell's Information Services, Livingston, N.J.

26. Elaine Landau has written a series of books for children that explore unusual anatomies. See, for example, *Joined at Birth: The Lives of Conjoined Twins* (New York: Franklin Watts, 1997).

27. See Joseph P. Shapiro, *No Pity: People with Disabilities Forging a New Civil Rights Movement* (New York: Three Rivers Press, 1993), pp. 54, 109. For a comparison of slavery and institutionalization, see ibid., pp. 159–160.

1. The Limits of Individuality

1. A common misconception is that P. T. Barnum thought up the phrase "Siamese twins" to designate Chang and Eng. For evidence that Chang and Eng themselves probably coined the term, see Irving Wallace and Amy Wallace, *The Two* (New York: Simon and Schuster, 1978), p. 74.

2. James Young Simpson, "A Lecture on the Siamese and Other Viable United Twins," *British Medical Journal*, 1 (1869): 139–141 and 229–233.

3. Wallace and Wallace, *The Two*, p. 22. When I told him the story of how Chang and Eng's mother had raised them, Dr. Jeffrey L. Marsh, a surgeon who treats children with craniofacial anomalies, responded: "Chang and Eng's mother exhibited the behavior that I learned long ago was the best predictor of behavioral outcome for an infant or young child with a facial birth defect (clinical observations only—no double-blind, randomized control study): if the parents were calm, accepting, and treated the child like their other offspring, neither being overly protective nor overly permissive, the child did well; if the parents never calmed down, the child had a high probability of having major psychosocial issues." Personal communication, May 2003.

4. Simpson, "A Lecture," p. 139.

5. On the question of how much she received, see Wallace and Wallace, *The Two*, p. 44.

6. Simpson, "A Lecture," p. 140.

7. Quoted in Wallace and Wallace, *The Two*, p. 173.

8. On this point, see Wallace and Wallace, *The Two*, p. 169. For a study of the stigma of obesity in American culture, see April Michelle Herndon, *Carrying the Torch: Fatness and Nation in the Age of Weight Loss* (Doctoral dissertation, Michigan State University, 2003).

9. Ibid., p. 215. According to Wallace and Wallace, the Bunkers first experimented with separate households in 1852.

10. Catalogued as "The Chang and Eng Bunker papers, 1833–1874; 1933–1967 [manuscript]," record number 3761, Southern Historical Collection, Wilson Library, University of North Carolina at Chapel Hill.

11. Quoted in Wallace and Wallace, *The Two*, p. 289.

12. William H. Pancoast, "Report on the Surgical Considerations in Regard to the Propriety of an Operation for the Separation of Eng and Chang Bunker, Commonly Known as the Siamese Twins," *Transactions of the College of Physicians* (Philadelphia), 1 (1875): 150.

13. The Wallaces claim that Chang and Eng did want to be separated but provide no evidence to back up this assertion—though they do provide evidence that challenges it (*The Two*, pp. 268, 280, etc.). They also claim that Chang and Eng considered separation in order to marry but that their future wives stopped them, fearing the operation would be deadly (pp. 175–176). Again, the Wallaces provide no documenta-

tion, and the statement seems contradicted by other claims and documents.

14. On the ways in which medical and scientific experts have been used by people with unusual anatomies to drum up profitable publicity, see Chapter 4. See also Alice Domurat Dreger, "Jarring Bodies: Thoughts on the Display of Unusual Anatomies," *Perspectives in Biology and Medicine*, 43 (Winter 2000): 161–172.

15. Pancoast, "Report," p. 156.

16. Ibid., p. 150.

17. Wallace and Wallace, *The Two*, p. 303.

18. See, for example, the letter from Brooklyn dated January 29, 1874, in the Chang and Eng Bunker Papers, University of North Carolina, Chapel Hill.

19. Wallace and Wallace, *The Two*, pp. 318–319.

20. The Bunker Papers at the University of North Carolina include an undated newspaper clipping, clearly published shortly after the deaths of Chang and Eng, which states: "It appears there is a provision in the wills of the twins against the severance of their bodies after death."

21. For images and discussion of the items in the museum's collection, see Gretchen Worden, *Mütter Museum of the College of Physicians of Philadelphia* (New York: Blast Books, 2002).

22. Jessie Bunker Bryant, *The Connected Bunkers* (Winston-Salem, N.C.: Jostens Graphics, 2002).

23. Margo Miles-Carney, personal communication, July 26, 2002, and November 15, 2002.

24. Quoted in Jimmy Tomlin, "Woman Compiles Book about Siamese Twins' Descendants," *High Point Enterprise* (High Point, N.C.), January 16, 2002.

25. On this history, see Iris Chang, *The Chinese in America: A Narrative History* (New York: Viking, 2003). For a hair-raising account of the display of an Eskimo man before and after his death—an exhibit mounted by the American Museum of Natural History in New York— see Kenn Harper, *Give Me My Father's Body: The Life of Minik, the New York Eskimo* (Frobisher Bay, N.W.T., Canada: Blacklead Books, 1986).

26. Rowena Spencer, *Conjoined Twins: Developmental Malformations and Clinical Implications* (Baltimore: Johns Hopkins University Press,

2003), presents the case for a theory of fusion—the idea that two embryos come together to form the conjoined twins. But in a scientifically and historically sophisticated review of Spencer (forthcoming in *Pediatric and Developmental Pathology*), J. Bruce Beckwith argues persuasively against some of her claims for fusion.

27. These figures come from Spencer, *Conjoined Twins*. Alternative figures are provided in L. Spitz and E. M. Kiely, "Experience in the Management of Conjoined Twins," *British Journal of Surgery*, 89 (2002): 1188–1192; Spitz and Kiely indicate that thoracopagus twins account for 40 percent of cases reported, omphalopagus for 32 percent, pygopagus for 19 percent, ischiopagus for 6 percent, and craniopagus for 2 percent. Spencer's sample is significantly larger, and therefore presumably more accurate.

28. Spencer, *Conjoined Twins*, ch. 12.

29. Ibid., p. 109.

30. Figures on viability are taken from M. C. Albert et al., "The Orthopedic Management of Conjoined Twins: A Review of Thirteen Cases and Report of Four Cases," *Journal of Pediatric Orthopaedics*, 12 (1992): 300. Spitz and Kiely write that "the incidence of conjoined twins is estimated at one in 50,000 pregnancies but, as around 60 per cent are stillborn, the true incidence is about one in 200,000 live births" ("Experience," p. 1188). Many present-day authors rely on the overall estimate given in L. D. Edmonds and P. M. Layde, "Conjoined Twins in the United States, 1970–1977," *Teratology*, 25 (1982): 301–308. Edmonds and Layde give an estimate of 10.25 cases per million births, a number reflecting the apparent incidence of conjoined twin births in the U.S. from 1970 to 1977; yet that figure may be unjustly low because it is based on clearly recorded incidences. Edmonds and Layde themselves speculate that "the true incidence of conjoined twinning most likely lies between the rate of about 1/100,000 live births . . . and the rate of 1/30,000" (p. 305). Many American and European authors believe the incidence to be lower among "Caucasians" than among other "racial" groups (see, e.g., Albert et al., "The Orthopedic Management," p. 300). It is difficult to know, however, the extent to which such statistics are now skewed by irregular reporting and by prenatal diagnosis and abortion.

31. J. Iveson-Iveson, "Siamese Twins," *Nursing Mirror*, 158 (1984): inside back cover.

32. A. E. Winch and M. T. Gonyea, "Separation of Conjoined Twins: A Case Study in Critical Care," *Critical Care Nursing Clinics of North America*, 6 (1994): 809.

33. Even within the disability rights movement—which encourages people to understand physical and moral interdependence as an integral aspect of all human life—many activists avoid any language that might cast doubt on the independent-individual status of a person with a disability. So programs that support people with disabilities living outside institutions are called "independent living centers"; and personal attendants who help with bodily upkeep are called "personal assistants" to indicate their status as mere employees, no matter how physically intimate the work. On the language of "attendant" versus "assistant," see Joseph P. Shapiro, *No Pity: People with Disabilities Forging a New Civil Rights Movement* (New York: Three Rivers Press, 1993), pp. 247–248. For a brief history of the Independent Living Movement, see ibid., ch. 2.

34. On this point, see Carl Elliott, *Better Than Well: American Medicine Meets the American Dream* (New York: W. W. Norton, 2003).

35. For alternative visions of human communities, see Alasdair MacIntyre, *Dependent Rational Animals: Why Human Beings Need the Virtues* (Chicago: Open Court, 1999).

36. Quoted in Marlene L. S. Cady, "My Siamese Twins Have Brought Me Joy," *Redbook*, February 1987, p. 32.

37. Marlene Cady, "The Pure Joy of Being Alive," *People Weekly*, July 3, 1989, pp. 64–71.

38. Cady, "My Siamese Twins," p. 32.

39. Ibid., p. 34. For a more detailed version of this story, see Cady, "The Pure Joy," p. 67.

40. Cady, "The Pure Joy," p. 67.

41. Ibid., p. 68.

42. Ibid., p. 70.

43. Ibid., p. 69.

44. Ibid., p. 70.

45. Ibid.

46. Ibid., p. 71.
47. K. Hubbard, "A Gift of Grace," *People*, 34 (1991): 44.
48. J. David Smith, *Psychological Profiles of Conjoined Twins: Heredity, Environment, and Identity* (New York: Praeger, 1988), p. 3.
49. Anonymous, "Siamese Twins Buried in Specially-Made Casket," *Jet*, 83 (1993): 16–17.
50. Smith, *Psychological Profiles*, p. 3.
51. Anonymous, "Siamese Twins Buried," p. 17.
52. Disability rights advocates have similarly noted that much more money is available to support people with disabilities living in institutions than to support those who seek to live in the general community.
53. Smith, *Psychological Profiles*, p. 4.
54. Anonymous, "Siamese Twins Buried," p. 16.
55. Smith, *Psychological Profiles*, p. 4.
56. Anonymous, "Siamese Twins Buried," p. 17.
57. Smith, *Psychological Profiles*, p. 60.
58. Anonymous, *Medical Description and Songs of Miss Millie-Christine, the Two-Headed Nightingale,* undated pamphlet in the collection of the State Historical Society of Wisconsin (Madison); microform number PAH-B-1583.c.1, Michigan State University Library, East Lansing, Michigan.
59. Smith, *Psychological Profiles*, p. 63.
60. Anonymous, *Medical Description*, p. 15.
61. K. Miller, "Together Forever," *Life*, April 1996, pp. 44–56.
62. C. Wallis, "The Most Intimate Bond," *Time*, 147 (1996): 64.
63. Miller, "Together Forever," p. 55.
64. Ibid., p. 49.
65. S. Begley, "Siamese Twins: From Ripley's to College," *Newsweek*, September 21, 1987, p. 72.
66. Smith, *Psychological Profiles*, p. 4.
67. T. Sweeting and P. Patterson, "Lin and Win Htut: The Conjoined Twins from Burma," *Canadian Nurse*, 80 (1984): 18.
68. Ibid., p. 20, emphasis added.
69. Pancoast, "Report," p. 156.
70. Hubbard, "A Gift," p. 43.

71. Pancoast, "Report," p. 157.

72. Smith, *Psychological Profiles*, p. 74.

73. On this point, for see S. E. Cleveland, E. Reitman, and D. Sheer, "Psychological Appraisal of Conjoined Twins," *Journal of Projective Techniques and Personality Assessment*, 28 (1964): 265–270. See also the case recounted in A. Pena Chavarria and P. G. Shipley, "The Siamese Twins of Espanola," *Annals of Medical History*, 6 (1924): 297–302. Concerning the Twins of Espanola, the authors write: "It was clearly seen that they were two individuals and had two souls and different minds" (p. 300). The Hungarian sisters Helen and Judith "differed from each other in character and temperament" (Simpson, "A Lecture," p. 230). Millie and Christina, the Carolina twins, were "in every sense somewhat different in dispositions and temperaments. . . . Their minds and mental acts were so separate and distinct, that one sometimes ate while the other was asleep" (ibid., p. 230). Rita and Christina, born in France in the early nineteenth century, "were distinctly dual in mind, though partially single in body. Often one head slept soundly, while the other head was awake and sucked. Sometimes, one cried while the other smiled" (ibid., p. 231). On Rita and Christina, see also Stephen Jay Gould, "Living with Connections," *Natural History*, 91 (1982): 18–22. A "double-monster" who lived in Scotland during the reign of James IV had, according to one report, "differing Passions, and diverse Wills" (history of Drummond as quoted in Simpson, "A Lecture," p. 231).

74. See Smith, *Psychological Profiles*.

75. Simpson, "A Lecture," p. 140.

76. H. Allen, "Report of an Autopsy on the Bodies of Chang and Eng Bunker, Commonly Known as the Siamese Twins," *Transactions of the College of Physicians* (Philadelphia), 1 (1875): 4.

77. B. H. Breakstone, "The Last Illness of the Blazek (Grown-Together) Twins," *American Medicine*, 17 (1922): 225.

78. Lori Schappell, personal communication, December 9, 2002.

79. Jane Mulkerrins, "Can These Siamese Twins Ever Live Apart?," *Daily Mail* (London), June 13, 2003, p. 13.

80. Erika Niedowski, "Adult Twins Risk All on Surgical Parting," *Chicago Tribune*, July 2, 2003, pp. 1, 16.

81. Mulkerrins, "Can These Siamese Twins Ever Live Apart?"

82. Anonymous, "Twins Will Be Split Next Month," *International Iran Times,* June 6, 2003, p. 1.

83. Mulkerrins, "Can These Siamese Twins Ever Live Apart?"

84. Ibid.

85. Ibid.

86. Quoted on *CBS Evening News,* July 8, 2003; transcript available from Burrelle's Information Services.

87. Quoted in Anonymous, "Nation in Shock over Death of Iranian Twins," *Belfast News Letter* (Northern Ireland), July 9, 2003, p. 14.

88. Ibid.

89. Wayne Arnold and Denise Grady, "Twins Die Trying to Live Two Lives," *New York Times,* July 9, 2003, p. A6.

90. See, for example, Daniel Henninger, "Iranian Twins Left the World a Big Idea," *Wall Street Journal,* July 11, 2003, p. A8.

91. See, for example, the editorial "Twins Succeeded in Living to Fullest," *Chicago Sun-Times,* July 9, 2003, p. 51.

92. See, for example, Cleveland, Reitman, and Sheer, "Psychological Appraisal," p. 269: "Jones et al. (1948) report that at age 34 the twins were offered the opportunity for separation and refused, explaining, 'It would be like an amputation of a personal appendage.'"

93. Simpson, "A Lecture," p. 141.

94. I do not mean to agree with this assessment, but merely wish to point out the inconsistent assumptions about what counts as a healthy body image.

95. K. Lipskey, "Conjoined Twins: Psychosocial Aspects," *AORN Journal,* 35 (1982): 58. For similar remarks, see Cleveland, Reitman, and Sheer, "Psychological Appraisal," p. 266; and see C. K. Pepper, "Ethical and Moral Considerations in the Separation of Conjoined Twins," *Birth Defects Original Article Series,* 3 (1967): 130.

96. See Nancy L. Segal, *Entwined Lives: Twins and What They Tell Us about Human Behavior* (New York: Plume, 1999).

97. Miller, "Together Forever," p. 56.

98. Cleveland, Reitman, and Sheer, "Psychological Appraisal," p. 268.

99. Ibid., p. 268.

100. Ibid. In the TAT (Thematic Apperception Test), subjects are shown a number of pictures and asked to describe the thoughts and feelings evoked in them by the images.
101. Ibid., p. 267.
102. Ibid.
103. Ibid., p. 269.
104. Jan Bondeson, "The Biddenden Maids: A Curious Chapter in the History of Conjoined Twins," *Journal of the Royal Society of Medicine,* 85 (1992): 217–221.
105. Anonymous, *Medical Description,* p. 25.
106. Quoted in Wallace and Wallace, *The Two,* p. 297.
107. Smith, *Psychological Profiles,* p. 77.
108. Interview with Ruta Sharangpani, November 14, 2003.
109. Camryn Manheim, *Wake Up, I'm Fat!* (New York: Broadway Books, 1999), p. 85. For an exploration of how fat functions as an identity, see Lisa Schoenfielder and Barb Wieser, eds., *Shadow on a Tightrope: Writings by Women on Fat Oppression* (San Francisco: Aunt Lute Books, 1983).
110. Joanne Green, "The Reality of the Miracle: What to Expect from the First Surgery," copyright 1996, available at widesmiles.org/cleftlinks/WS-162.html.
111. Another reason may be the guilt that people, particularly parents, feel after life-altering illnesses or accidents, though guilt is hardly an unknown emotion to parents of children born with unusual anatomies. On the history of maternal guilt, influence, and power, see Marie-Hélène Huet, *Monstrous Imagination* (Cambridge, Mass.: Harvard University Press, 1993); and Harriet Ritvo, *The Platypus and the Mermaid and Other Figments of the Classifying Imagination* (Cambridge, Mass.: Harvard University Press, 1997), pp. 112–113. For a sensitive and intelligent discussion of how identity and relationships are formed through narrative, and how traumatic body changes disrupt or alter life narratives (and therefore identities and relationships), see Howard Brody, *Stories of Sickness,* 2nd ed. (New York: Oxford University Press, 2003), esp. pp. 245–247. People who become disabled long after birth may be less tolerant of discrimination than those born with

stigmatized anatomies, for the very reasons Brody articulates. They see "no reason suddenly to forfeit the first-class citizenship they had grown up to expect as a birthright" (Shapiro, *No Pity,* p. 144).

112. Lori Schappell, personal communication, December 9, 2002. I am grateful to Adrienne Asch for reminding me not to imply that anatomical difference necessarily produces a given identity, experience, or meaning.

113. James A. Rooth, "The Brighton United Twins," *British Medical Journal,* 2 (1911): 653–654.

114. Smith, *Psychological Profiles,* p. 69 (citing F. Drimmer).

115. Ibid., pp. 69–71.

116. Ibid., p. 74.

117. Ricky Jay, "Sisters, United: Step Right Up!" *New York Times,* October 12, 1997, p AR5. On this point, see also Smith, *Psychological Profiles,* p. 74.

118. Harry Haun, "Side by Side by Side Show," *Playbill* for the Broadway musical *Sideshow,* October 1997.

119. *Face to Face: The Story of the Schappell Twins,* a film by Ellen Weissbrod, *A&E Biography* (2000).

120. "I've Got You under My Skin," music and lyrics by Cole Porter, 1936; copyright © The Cole Porter Trusts, New York, N.Y.

2. Split Decisions

1. Theophanes Continuatus, quoted in G. E. Pentogalos and J. G. Lascaratos, "A Surgical Operation Performed on Siamese Twins During the Tenth Century in Byzantium," *Bulletin of the History of Medicine,* 58 (1984): 99.

2. For a similar fifteenth-century case, see J. L. Calagan, "Conjoined Twins Born near Worms, 1495," *Journal of the History of Medicine and Allied Sciences,* 38 (1983): 450–451.

3. James Y. Simpson, "A Lecture on the Siamese and Other Viable United Twins," *British Medical Journal,* 1 (February 13 and March 13, 1869): 141.

4. For a catalog of twenty-five separation attempts from 1689 to 1963,

see W. B. Kiesewetter, "Surgery on Conjoined (Siamese) Twins," *Surgery*, 59 (1966): 860–871.

5. Ibid., p. 861.

6. Mark Stringer, personal communication, November 3–5, 2002.

7. Quotation from the film *Katie and Eilish*, by Mark Galloway (Yorkshire Television, 1992), as noted in Catherine Myser and David L. Clark, "'Fixing' Katie and Eilish: Medical Documentaries and the Subjection of Conjoined Twins," *Literature and Medicine*, 17 (Spring 1998): 54.

8. Ibid., p. 54, referring to Galloway's film *Katie and Eilish*.

9. Patrick Deasy as quoted in Myser and Clark, "'Fixing' Katie and Eilish," p. 54.

10. L. Spitz, M. D. Stringer, E. M. Kiely, P. G. Ransley, and P. Smith, "Separation of Brachio-Thoraco-Omphalo-Ischiopagus Bipus Conjoined Twins," *Journal of Pediatric Surgery*, 29 (April, 1994): 477.

11. Ibid.

12. Ibid., pp. 477–478.

13. Ibid., p. 477.

14. Mark Stringer, personal communication, November 3–5, 2002.

15. Ibid.

16. Ibid. For a discussion of the possible reasons for Katie's death, see also Spitz et al., "Separation," p. 479–480.

17. Mark Stringer, personal communication, November 3–5, 2002.

18. Spitz et al., "Separation," p. 479.

19. Myser and Clark, "'Fixing' Katie and Eilish," p. 63.

20. On why being "normal" is not necessarily the same as being "healthy," see Phillip V. Davis and John G. Bradley, "The Meaning of Normal," *Perspectives in Biology and Medicine*, 40, no. 1 (Autumn 1996): 68–76.

21. On this point, see Joan Ablon, "Ambiguity and Difference: Families with Dwarf Children," *Social Science and Medicine*, 30, no. 8 (1990): 879–887.

22. See Arthur W. Frank, *The Wounded Storyteller: Body, Illness, and Ethics* (Chicago: University of Chicago Press, 1995), ch. 4. Also see Alice Domurat Dreger, *Hermaphrodites and the Medical Invention of Sex* (Cambridge, Mass.: Harvard University Press, 1998), pp. 184–186.

23. More generally, F. D. Reeve writes that when an individual's identity is changed by the onset of a disability, relatives' and friends' identities also change. See Reeve, "Relatively Disabled," *Michigan Quarterly Review*, 37, no. 3 (Summer 1998): 541–546.

24. Monica J. Casper discusses the way in which mothers who choose fetal surgery are praised as heroes. See Casper, *The Making of the Unborn Patient: A Social Anatomy of Fetal Surgery* (New Brunswick, N.J.: Rutgers University Press, 1998), pp. 180–182.

25. J. Douglas, "Double Miracle," *Nursing Times,* 85 (1989): 17.

26. Simon Mawer, *Mendel's Dwarf* (New York: Harmony Books, 1998), p. 5.

27. In *Autobiography of a Face* (Boston: Houghton Mifflin, 1994), Lucy Grealy describes her efforts to cope with torturous cancer treatments. Her mother's insistence that she be "brave," she says, both silenced her and made her feel like a failure for experiencing pain and fear.

28. Sometimes there can be little hope of social accommodation to functional challenges. In 2002 the Guatemalan twins Maria de Jesus and Maria Teresa Quiej-Alvarez were separated, and their mobility was an important goal. The girls had been born joined at the top of the head and facing in different directions, making it highly unlikely they would have achieved substantial mobility. If left conjoined they might well have suffered bedsores and other physiological pathologies arising from immobility, and they would have had a much harder time than twins like Katie and Eilish getting out and about in the world; they would probably have had to stay prone, on a double-length stretcher. But this situation is quite extraordinary, even among conjoined twins.

29. It is not unusual, however—just before or just after a separation surgery—for anxious surgeons or the press to announce some new physiological finding that supposedly has made separation necessary to preserve health or life. In the case of Ladan and Laleh Bijani, for example, the surgical team announced that just prior to separation they had found out the patients' intracranial pressure had risen to dangerous levels. But in this case, as in most others, options for treating the physiological concern were not limited to separation; separation was substantially more dangerous than the alternatives, yet the alternatives

were not pursued. This is because the main goal of these surgeries is to normalize appearance and produce physical independence.

30. Sometimes surgeons seem to take comfort in the "success" of separation even when patients die. After Laleh and Ladan Bijani died during separation, the lead surgeon, Keith Goh told the press, "At least we helped them achieve their dream of separation." Quoted in Anonymous, "Nation in Shock over Death of Iranian Twins," *Belfast News Letter* (Northern Ireland), July 9, 2003, p. 14.

31. Ladan and Laleh Bijani may have sought separation to increase the likelihood that each would become a wife and mother. They did not name this publicly as a motivation but, being Iranian Islamic women, they may have been hesitant to allude to their romantic and sexual lives.

32. Quoted in Kenneth Miller, "Together Forever," *Life,* April 1996.

33. See Catherine L. Minto et al., "The Effect of Clitoral Surgery on Sexual Outcome in Individuals Who Have Intersex Conditions with Ambiguous Genitalia: A Cross-Sectional Study," *Lancet,* 361 (April 12, 2003): 1252–1257.

34. Janet Kornblum, "'Sublime Leadership Made Surgery Work': Meet the Man Who Separated the Twins," *USA Today,* August 8, 2002.

35. William H. Pancoast, "Report on the Surgical Considerations in Regard to the Propriety of an Operation for the Separation of Eng and Chang Bunker, Commonly Known as the Siamese Twins," *Transactions of the College of Physicians* (Philadelphia), 1 (1875): 156.

36. Ibid., p. 154.

37. Quoted in Anonymous, *Medical Description and Songs of Miss Millie-Christine, the Two-Headed Nightingale,* undated pamphlet in the collection of the State Historical Society of Wisconsin (Madison); microform number PAH-B-1583.c.1, Michigan State University Library, East Lansing, Michigan, p. 25.

38. Quoted in J. David Smith, *Psychological Profiles of Conjoined Twins: Heredity, Environment, and Identity* (New York: Praeger, 1988), p. 72.

39. G. A. McLorie, A. E. Khoury, and T. Alphin, "Ischiopagus Twins: An Outcome Analysis of Urological Aspects of Repair in Three Sets of Twins," *Journal of Urology,* 157, no. 2 (February 1997): 650–653.

40. S. Cywes et al., "Conjoined Twins: The Cape Town Experience," *Pedi-*

atric Surgery International, 12, no. 4 (April 1997): 234–248. Compare this article with J. A. O'Neill, Jr., et al., "Surgical Experience with Thirteen Conjoined Twins," *Annals of Surgery,* 208, no. 3 (September 1988): 299–312.

41. See Spitz and Kiely, "Experience."

42. I am indebted to Libby Bogdan-Lovis for this concept and term.

43. Mark Stringer, personal communication, November 3–5, 2002.

44. I am grateful to Mark Stringer for this insight.

45. Mark Stringer, personal communication, November 3–5, 2002.

46. See the interview with Benjamin Carson on *Nightline,* July 11, 2003; transcripts available from ABC News Transcripts.

47. Rowena Spencer, *Conjoined Twins: Developmental Malformations and Clinical Implications* (Baltimore, Md.: Johns Hopkins University Press, 2003), pp. 310–311.

48. See, for example, John Money, *Hermaphroditism: An Inquiry into the Nature of a Human Paradox* (Doctoral dissertation, Harvard University, 1952); and E. Clifford, "Why Are They So Normal?" *Cleft Palate Journal,* 20, no. 1 (January 1983): 83–84.

49. Quoted in Anonymous, "50 Percent Chance of Both Twins Surviving Split," *International Iran Times,* June 20, 2003, p. 3.

50. For a critique of the idea that female genitalia are harder to construct than male, see Alice Domurat Dreger, "'Ambiguous Sex'—or Ambivalent Medicine? Ethical Problems in the Treatment of Intersexuality," *Hastings Center Report,* 28 (May–June 1998): 24–35.

51. Spitz and Kiely, "Experience."

52. O'Neill, Jr., "Surgical Experience," p. 303.

53. J. Savickis, "The Separation of Conjoined Twins: An OR Nursing Perspective," *Canadian Nurse,* 80 (December 1984): 21–23.

54. Anonymous, "Nepalese Twins Return Home," *BBC News,* December 18, 2001 (news.bbc.co.uk/1/hi/world/asia-pacific/1662777.stm).

55. A. E. Winch and M. T. Gonyea, "Separation of Conjoined Twins: A Case Study in Critical Care," *Critical Care Nursing Clinics of North America,* 6 (1994): 808.

56. On the trauma experienced by one twin upon the death of the other, see Nancy L. Segal, *Entwined Lives: Twins and What They Tell Us about Human Behavior* (New York: Plume, 1999).

57. See Richard Lansdown et al., *Visibly Different: Coping with Disfigurement* (Oxford: Oxford University Press, 1997). On the value of support groups for parents and families, see Joan Ablon, "The Parents' Auxiliary of Little People of America: A Self-Help Model of Social Support for Families of Short-Statured Children," *Prevention in Human Services,* 1 (Spring 1982): 31–46. On the "natural acceptance of the disorder" of achondroplasia and the need to "direct the reaction of the family-surroundings," especially when normalizing limb-lengthening surgery is offered, see L. Ancona, "The Psychodynamics of Achondroplasia," *Basic Life Sciences,* 48 (1988): 447–451.

58. See Wendy Mouradian, "Deficits versus Strengths: Ethics and Implications for Clinical Practice and Research," *Cleft Palate Craniofacial Journal,* 38, no. 3 (May 2001): 255–559.

59. P. C. Thomas, "Multidisciplinary Care of the Child Born with Cleft Lip and Palate," *ORL Head Neck Nursing,* 18, no. 4 (Fall 2000): 6–16.

60. See, for example, Elizabeth S. McCaughey et al., "Randomised Trial of Growth Hormone in Short Normal Girls," *Lancet,* 351 (March 28, 1988): 940–944. See also the critique in Alice Dreger, "When Medicine Goes Too Far in the Pursuit of Normality," *New York Times,* July 28, 1998, p. B-10.

61. Edmund G. Howe, "Intersexuality: What Should Careproviders Do Now," *Journal of Clinical Ethics,* 9, no. 4 (Winter 1988): 337–344.

62. I suppose some would argue that this happens occasionally in certain supposed cases of ADD or ADHD in children. And critics of medicalized birth sometimes suggest that laboring women are medicated for pain relief because their cries disturb those around them.

63. For example, significant success in treating women with the intersex condition MRKH has been achieved with a support group. See P. T. Weijenborg and M. M. ter Kuile, "The Effect of a Group Programme on Women with the Mayer-Rokitansky-Kuster-Hauser Syndrome," *British Journal of Gynecology,* 107, no. 3 (March 2000): 365–368.

64. "First Do No Harm: Total Patient Care for Intersex," videotape, Intersex Society of North America (Seattle), 2002.

65. Sarah Creighton and Catherine Minto, "Managing Intersex," *British Medical Journal,* 323 (December 1, 2001): 1264–1265.

66. For example, the anti-circumcision movement seeks to put boys' right

to physical self-determination over some parents' desire to provide a "normal-looking" (i.e., circumcised) penis.

67. Priscilla Alderson, *Children's Consent to Surgery* (Buckingham, England: Open University Press, 1993).

68. American Academy of Pediatrics, "Informed Consent, Parental Permission, and Assent in Pediatric Practice (RE9510)," *Pediatrics*, 95, no. 2 (February 1995): 314–317.

69. Though the surgery was chosen by the patients themselves, the separation of Laleh and Ladan Bijani looked like a study in conflict of interests. Singapore's Raffles Hospital, where the operation took place, is a private, for-profit hospital that used its website both to publicize the case and to encourage private investment. Raffles' "stock rose 18 percent . . . after the operation on the twins began and appeared to be going smoothly." But it fell substantially when the twins' deaths were announced. See Anonymous, "Was It the Right Decision to Split," *International Iran Times*, July 11, 2003, p. 3.

70. See, for example, Kornblum, "'Sublime Leadership.'" For another good example of virtual hagiography, see Jane Black, "Making Broken Children Normal" (on surgeon Kenneth Salyer), *Business Week Online* (July 30, 2002): 1–3.

71. For a thoughtful antidote to the ideal of "objectivity" in medicine, see Jodi Halpern, *From Detached Concern to Empathy: Humanizing Medical Practice* (New York: Oxford University Press, 2001).

72. For example, an untraceable story long circulating among intersex clinicians tells of a child who was allowed to grow up with intersexed genitals and who killed himself later in life as a result of his condition. The way in which fear of the worst-case scenario drives doctors to pursue unnecessary and dangerous interventions is explored well in Howard Brody and J. R. Thompson, "The Maximin Strategy in Modern Obstetrics," *Journal of Family Practice*, 12, no. 6 (June 1981): 977–986.

73. Barron H. Lerner traces a similar ethos of intervention in *The Breast Cancer Wars: Hope, Fear, and the Pursuit of a Cure in Twentieth-Century America* (New York: Oxford University Press, 2001).

74. Adrienne Asch, "Distracted by Disability: The 'Difference' of Disabil-

ity in the Medical Setting," *Cambridge Quarterly of Healthcare Ethics*, 7 (1998): 80.

75. A. F. Guttmacher, "Biographical Notes on Some Famous Conjoined Twins," *Birth Defects Original Article Series*, 3 (1967): 10.

76. E. S. Golladay et al., "Dicephalus Dipus Conjoined Twins: A Surgical Separation and Review of Previously Reported Cases," *Journal of Pediatric Surgery*, 17 (1982): 263, emphasis added.

77. Valerie Miké, "Outcomes Research and the Quality of Health Care: The Beacon of an Ethics of Evidence," *Evaluation of the Health Professions*, 22 (1999): 10.

3. What Sacrifice

1. E. S. Golladay et al., "Dicephalus Dipus Conjoined Twins: A Surgical Separation and Review of Previously Reported Cases," *Journal of Pediatric Surgery*, 17 (1982): 260.

2. On the 1955 case, see W. B. Kiesewetter, "Surgery on Conjoined (Siamese) Twins," *Surgery*, 59 (1966): 860–871. By 1994, there had been at least six sacrifice surgeries, according to C. T. Chiu et al., "Separation of Thoracopagus Conjoined Twins: A Case Report," *Journal of Cardiovascular Surgery*, 35 (1994): 459–462. By 1996, there had been at least nine, according to D. C. Thomasma et al., "The Ethics of Caring for Conjoined Twins: The Lakeberg Twins," *Hastings Center Report*, 24 (1996): 4–12. Since then, sacrifice surgeries have also been performed in the cases of the Soto and Attard twins (discussed below), bringing the total to at least eleven.

3. George J. Annas, "Siamese Twins: Killing One to Save the Other," *Hastings Center Report*, 17 (1987): 27; and Golladay et al., "Dicephalus Dipus," p. 259.

4. See C. K. Pepper, "Ethical and Moral Considerations in the Separation of Conjoined Twins," *Birth Defects Original Article Series*, 3 (1967): 128–134.

5. See, for example, James A. O'Neill, Jr., et al., "Surgical Experience with Thirteen Conjoined Twins," *Annals of Surgery*, 208 (1988): 308; and Errol R. Norwitz et al., "Separation of Conjoined Twins with the Twin

Reversed-Arterial-Perfusion Sequence after Prenatal Planning with Three-Dimensional Modeling," *New England Journal of Medicine,* 343 (2000): 400.

6. Quoted in Thomasma et al., "The Ethics of Caring," p. 5.

7. Quoted in Anonymous, "Siamese Twin Op Details Revealed," *BBC News,* December 7, 2000.

8. Ibid.

9. See A. E. Winch and M. T. Gonyea, "Separation of Conjoined Twins: A Case Study in Critical Care," *Critical Care Nursing Clinics of North America,* 6 (1994): 809.

10. O'Neill et al., "Surgical Experience," p. 308. Compare the value-laden comments of Adrian Bianchi about the Attard girls: "The blood was going from the good twin to the—to the poorer twin." Quoted in "A Family's Faith," *ABC News Prime Time,* December 14, 2000 (online transcript).

11. Annas, "Siamese Twins," p. 29.

12. O'Neill et al., "Surgical Experience," p. 308.

13. The formal definition of "parasitic" twins has remained fairly stable since the teratologist Isidore Geoffroy Saint-Hilaire discussed the topic in the 1830s. Geoffroy described such twins as "inert, irregular masses, composed principally of bone, teeth, hair, and fat." See Isidore Geoffroy Saint-Hilaire, *Histoire générale et particulière des anomalies de l'organisation . . . ou Traité de tératologie,* vol. 2 (Paris: J.-B. Baillière, 1832–1836), p. 185.

14. R. Spencer, "Conjoined Twins: Theoretical Embryological Basis," *Teratology,* 45 (1992): 591.

15. See, for example, R. Drut, C. Garcia, and R. M. Drut, "Poorly Organized Parasitic Conjoined Twins: Report of Four Cases," *Pediatric Pathology,* 12 (1992): 691–700. See also Rowena Spencer, *Conjoined Twins: Developmental Malformations and Clinical Implications* (Baltimore, Md.: Johns Hopkins University Press, 2003), ch. 12.

16. J. J. Paris, "Ethical Issues in Separation of the Lakeberg Siamese Twins," *Journal of Perinatology,* 13 (1993): 423.

17. Jean Seligmann, "Is It More Humane Not to Operate?" *Newsweek,* August 23, 1993, p. 44.

18. Thomasma et al., p. 4.

19. Ibid., p. 5.

20. Ibid.

21. Ibid.

22. Jonathan Muraskas, Letter to the Editor, *Journal of Perinatology*, 14 (1994): 168.

23. Ibid.

24. Thomasma et al., p. 5.

25. See the comments by Ken Lakeberg in Anastasia Toufexis, "The Ultimate Choice," *Time*, August 30, 1993, p. 43.

26. Stephen E. Lammers, "The Lakeberg Case: Tragedies and Medical Choices," *Christian Century*, 110 (1993): 845. For a direct quote from Dr. O'Neill on this point, see Toufexis, "The Ultimate Choice," p. 44.

27. Thomasma et al., p. 8.

28. See, for example, Stephen J. Gould, "Living with Connections," *Natural History*, 91 (1982): 18–22.

29. Toufexis, "The Ultimate Choice," p. 44.

30. Anastasia Toufexis, "The Brief Life of Angela Lakeberg," *Time*, June 27, 1994, p. 61.

31. Winch and Gonyea, "Separation of Conjoined Twins," p. 809.

32. Thomasma et al., p. 11.

33. Jill Porter, "'Sweetums' Was a Riot when You Got Her Going," *Philadelphia Daily News*, June 10, 1994, p. 5.

34. Toufexis, "The Brief Life," p. 62.

35. Porter, "'Sweetums' Was a Riot."

36. Toufexis, "The Brief Life," p. 62. See also Ron Goldwyn and Dave Bittan, "The Lakeberg Twins: Infection Kills Surviving Sister at Age 11 Months," *Philadelphia Daily News*, June 10, 1994, pp. 4, 39.

37. Quoted in Anonymous, "Twin Who Survived Separation Surgery Dies," *New York Times*, June 10, 1994, p. A14.

38. Ibid.

39. Ronald Dworkin, "Is High-Tech Life-Saving Noble or Simply a Waste?" *International Herald Tribune*, September 1, 1993, p. 7.

40. Lammers, "The Lakeberg Case," p. 846.

41. Quoted in Becky Batcha, "The Legacy: A Critical Look at Such Surgery," *Philadelphia Daily News*, June 10, 1994, p. 5.

42. Muraskas, Letter to the Editor, p. 168.

43. Alan R. Fleischman, Letter to the Editor, *Journal of Perinatology,* 14 (1994): 169.

44. Ibid.

45. Seligmann, "Is It More Humane Not to Operate?" p. 44.

46. Lammers, "The Lakeberg Case," p. 845.

47. Prior to the Lakeberg separation, the ethics of sacrifice surgery had been discussed in two articles: Annas, "Siamese Twins," and Pepper, "Ethical and Moral Considerations."

48. Annas, "Siamese Twins," p. 28.

49. Thomasma et al., p. 8.

50. Ibid., p. 9.

51. A court gag order (which was later lifted) required that the girls' identities be concealed: in the trial records and in most of the press accounts of this case, Gracie was known as Jodie, and Rosie was known as Mary. I have used their real names here to avoid confusion.

52. For an analysis of this case from a similar point of view (one influenced by my previous work), see Y. Michael Barilan, "Head-Counting vs. Heart-Counting: An Examination of the Recent Case of the Conjoined Twins from Malta," *Perspectives in Biology and Medicine,* 45, no. 4 (Autumn 2002): 593–605.

53. Voiceover comments of Charles Gibson, "A Family's Faith."

54. Comments of Rina Attard in "A Family's Faith."

55. See the decision by Justice Johnson in the High Court Justice Family Decision, case no. FD00P10893, August 25, 2000; and see Anonymous, "Jodie and Mary: The Medical Facts," *BBC News,* December 7, 2000 (online).

56. Anonymous, "Siamese Twin Separation 'Lawful,'" *BBC News,* September 13, 2000 (online).

57. Quoted in Anonymous, "Siamese Twins: A Surgeon's View," *BBC News,* September 22, 2000 (online).

58. John Arlidge, "Twins' Parents Seek a Quarter Million Pounds for Story," *Guardian Limited,* September 24, 2000 (online).

59. Comments of Rina Attard in "A Family's Faith."

60. Adrian Whitfield, Queen's Counsel, quoted in Anonymous, "Siamese Twin Separation 'Lawful.'"

61. Quoted in Johnson, "Decision."

62. Johnson, "Decision."

63. Ibid.

64. Ibid.

65. Ibid.

66. Ibid.

67. George J. Annas, "Conjoined Twins: The Limits of Law at the Limits of Life," *New England Journal of Medicine,* 344 (2001): 1104.

68. Anonymous, "Vatican 'Haven' for Siamese Twins," *CNN.com,* August 28, 2000 (online).

69. Anonymous, "Siamese Twin Separation 'Lawful.'"

70. Quoted ibid.

71. Anonymous, "Conjoined Twins: Judge Calls for Second Opinion," *CNN.com,* September 5, 2000 (online).

72. Anonymous, "Siamese Twin Separation 'Lawful.'"

73. Lord Justice Ward, "Re A (children)," (2000), 4 All ER 961.

74. Ibid.

75. Annas, "Conjoined Twins," p. 1104.

76. Lord Justice Brooke, "Re A (children)," (2000), 4 All ER 961.

77. Ibid.

78. Lord Justice Robert Walker, "Re A (children)," (2000), 4 All ER 961.

79. Ward, "Re A (children)."

80. Brooke, "Re A (children)."

81. Annas, "Conjoined Twins," p. 1105.

82. Ibid., p. 1106.

83. Ibid.

84. Ibid., p. 1104.

85. Quoted ibid., p. 1105 (emphasis added).

86. Ibid., p. 1106.

87. Ibid., p. 1105.

88. Quoted in Anonymous, "Ethics Expert: Twin Decision Wrong," *BBC News,* September 22, 2000 (online).

89. Quoted in Anonymous, "Siamese Twin Dies after Separation," *BBC News,* November 7, 2000 (online.)

90. John L. Allen, Jr., "Sophie's Choice: Conjoined Twins Give Birth to Moral and Legal Debate," *Second Opinion*(December 2000): 26.

91. Anonymous, "Bid to Halt Twins' Separation Fails," *CNN.com*, November 3, 2000 (online).

92. Anonymous, "Siamese Twin Op Details Revealed."

93. Quoted in Franco Aloisio, "Siamese Twins' Surgeon Speaks of Life and Death," *Malta Independent on Sunday*, December 17, 2000 (online).

94. Anonymous, "Siamese Twin Dies after Separation."

95. Anonymous, "Siamese Twin Mary Laid to Rest," *BBC News*, January 19, 2001 (online).

96. Anonymous, "Jodie's Parents Tell of Grief," *BBC News*, December 7, 2000 (online).

97. Comments of Bianchi in "A Family's Faith."

98. Quoted in Anonymous, "Long Road to Recovery for Twin Jodie," *BBC News*, November 7, 2000 (online).

99. Quoted in Anonymous, "Siamese Twin Returns Home," *BBC News*, June 17, 2001 (online).

100. Anonymous, "Jodie and Mary: The Medical Facts."

101. Anonymous, "Siamese Twin Op Details Revealed."

102. Quoted in Anonymous, "Siamese Twin Returns Home."

103. Anonymous, "Jodie's Parents Tell of Grief."

104. Anonymous, "Jodie and Mary: The Medical Facts."

105. Allen, "Sophie's Choice."

106. Anonymous, "Jodie's Parents Tell of Grief." Rina Attard repeated the claim in Anonymous, "Siamese Twin Laid to Rest."

107. Clare Dyer, "Judges Choose 'Lesser of Two Evils': Past Cases Little Help as Court Decides on Unique Case," *Guardian Unlimited*, September 23, 2000 (online).

108. Denise Grady, "A 'Miracle' Saves One of Conjoined Twins Who Shared a Heart," *New York Times*, August 11, 2000, p. 1.

109. Norwitz et al., "Separation of Conjoined Twins."

110. Grady, "A 'Miracle.'"

111. Ibid.

112. Denise Grady, "Fighting for Life: A Couple's Determination to Save Their Horribly Deformed Babies Turns into a Medical Tragedy," *Gazette* (Montreal), August 19, 2000, p. J6.

113. Norwitz et al., "Separation of Conjoined Twins," p. 400.

114. Ibid., p. 400.

115. Grady, "Fighting for Life."

116. Grady, "A 'Miracle.'"

117. Grady, "Fighting for Life."

4. Freeing the Irish Giant

1. *Liebe Perla,* directed by Shahar Rozen, produced by Edna Kowarksy (Israel and Germany: Eden Productions, 1999). Documentary, 63 minutes.

2. Sander L. Gilman has written extensively on the representation of "defectives," including Jews, in the history of medicine. See, for example, Gilman, *Sexuality: An Illustrated History, Representing the Sexual in Medicine and Culture from the Middle Ages to the Age of AIDS* (New York: Wiley, 1989); idem, *Difference and Pathology: Stereotypes of Sexuality, Race, and Madness* (Ithaca, N.Y.: Cornell University Press, 1985); idem, *Health and Illness: Images of Difference* (London: Reaktion Books, 1995).

3. Jan Bondeson and Elizabeth Allen, "Craniopagus Parasiticus: Everard Home's Two-Headed Boy of Bengal and Some Other Cases," *Surgical Neurology,* 31 (1989): 426–434.

4. Jessie Dobson, *Descriptive Catalogue of the Physiological Series in the Hunterian Museum of the Royal College of Surgeons of England,* Part II (London: E. and S. Livingstone, 1971), pp. 199–206.

5. Quoted ibid., pp. 200–201.

6. Ibid., p. 202.

7. For an exploration of the history, sociology, and politics of "freak" shows and related displays, see Rosemarie Garland Thomson, ed., *Freakery: Cultural Spectacles of the Extraordinary Body* (New York: New York University Press, 1996).

8. On this point, see Adrienne Asch, "Distracted by Disability: The 'Difference' of Disability in the Medical Setting," *Cambridge Quarterly of Healthcare Ethics,* 7 (1998): 83–84.

9. See Robert N. Proctor, *Racial Hygiene: Medicine under the Nazis* (Cambridge, Mass.: Harvard University Press, 1988), pp. 97–101.

10. See *A Little History Worth Knowing: Disability Down through the Ages* (Irene M. Woods and Associates, 1998). Video, 22 minutes. Distrib-

uted by Program Development Associates, Syracuse, N.Y. (www.pdassoc.com).

11. Kim [no last name], "As Is," in Alice Domurat Dreger, ed., *Intersex in the Age of Ethics* (Frederick, Md.: University Publishing Group, 1999), p. 99.

12. Quoted in Dobson, *Descriptive Catalogue,* p. 200.

13. William H. Pancoast, "Report on the Surgical Considerations in Regard to the Propriety of an Operation for the Separation of Eng and Chang Bunker, Commonly Known as the Siamese Twins," *Transactions of the College of Physicians* (Philadelphia), 1 (1875): 149–169.

14. Anonymous, *Medical Description and Songs of Miss Millie-Christine, the Two-Headed Nightingale,* undated pamphlet in the collection of the State Historical Society of Wisconsin (Madison); microform number PAH-B-1583.c.1, Michigan State University Library, East Lansing, Michigan. Emphasis added.

15. Ibid.

16. Ibid.

17. Robert Bogdan, *Freak Show: Presenting Human Oddities for Amusement and Profit* (Chicago: University of Chicago Press, 1988).

18. From my point of view, this is particularly the case with the specimen known as "Kennewick Man."

19. Audrey N. Bell, "Separating Conjoined Twins: A Care Plan," *AORN Journal,* 35 (1982): 53.

20. Catherine Myser and David L. Clark, "'Fixing' Katie and Eilish: Medical Documentaries and the Subjection of Conjoined Twins," *Literature and Medicine,* 17 (Spring 1998): 46.

21. On the "quest narrative" in medicine, see Arthur W. Frank, *The Wounded Storyteller: Body, Illness, and Ethics* (Chicago: University of Chicago Press, 1995), ch. 6.

22. Quoted in Myser and Clark, "'Fixing' Katie and Eilish," p. 62.

23. Kathi Wolfe wryly remarks, "My disability isn't a burden; having to be so damned inspirational is." Wolfe, "Ordinary People: Why the Disabled Aren't So Different," *Humanist,* 56 (November–December 1996): 31–34. Wolfe also quotes David Hevey, who says: "The non-disabled think either we can do nothing or that we can perform the most superhuman feats" (ibid., p. 31).

24. For an excellent critique of the practice of publishing photographs of patients in medical texts, see Sarah Creighton et al., "Medical Photography: Ethics, Consent, and the Intersex Patient," *BJU International,* 89 (2002): 67–72.

25. L. Spitz, M. D. Stringer, E. M. Kiely, P. G. Ransley, and P. Smith, "Separation of Brachio-Thoraco-Omphalo-Ischiopagus Bipus Conjoined Twins," *Journal of Pediatric Surgery,* 29 (April 1994): 477–481.

26. Cheryl Chase, personal communication, November 25, 2002.

27. Creighton et al. ("Medical Photography") discuss how often this happens, and imply that authors should publish with the expectation that their subjects will someday see the publications. This revelation led me to change my own use of medicalized photos of people who might be living. See Alice Domurat Dreger, "Avoiding the Fetal Position," *Studies in History and Philosophy of the Biological and Biomedical Sciences,* 30, no. 2 (1999): 255–261.

28. For an examination of the way in which depersonalized, seemingly objective language contributed to the abuses of the Tuskegee Syphilis Study, see Martha Solomon Watson, "The Rhetoric of Dehumanization: An Analysis of Medical Reports of the Tuskegee Syphilis Project," in Susan M. Reverby, ed., *Tuskegee's Truths: Rethinking the Tuskegee Syphilis Study* (Chapel Hill: University of North Carolina Press, 2000), pp. 251–265.

29. See, for example, Sherri Groveman's chapter in G. Michael Besser and Michael O. Thorner, eds., *Comprehensive Clinical Endocrinology,* 3rd ed. (St. Louis: Mosby, 2002).

30. *Dwarfs: Not a Fairytale,* a film by Lisa Abelow Hedley (2000), a project of the Children of Difference Foundation.

31. Asch, "Distracted," p. 78.

32. Ibid., p. 77.

33. For other examples of artists challenging traditional narratives of disease, deformity, and disability, see Diane Kirkpatrick, "Images of Disability," *Michigan Quarterly Review,* 37, no. 3 (Summer 1998): 426–440.

34. *Face to Face: The Story of the Schappell Twins,* directed by Ellen Weissbrod (A&E Television, 2000). Documentary, 100 minutes.

35. G. Thomas Couser, "Double Exposure: Performing Conjoined Twinship," unpublished paper, 2002. Quoted with permission.

36. Ibid.
37. Ibid.
38. Ibid.
39. Ibid.
40. For a fascinating and subversive comparative history of freak shows and beauty pageants, see Rosemarie Garland Thomson, "The Beauty and the Freak," *Michigan Quarterly Review,* 37, no. 3 (Summer 1998): 459–474.
41. Danny Black, personal communication, December 13, 2002.
42. For an example of how even visibly disabled people are essentially forced into closets of silence, and how radical representations of them can function as "coming out" experiences, see Jim Ferris, "Uncovery to Recovery: Reclaiming One Man's Body on a Nude Photo Shoot," *Michigan Quarterly Review,* 39, no. 3 (Summer 1998): 503–518. See also Thomson, *Freakery,* p. xvii.
43. J. Bruce Beckwith to Alice Dreger, May 13, 2000.
44. Alice Dreger to J. Bruce Beckwith, May 28, 2000.
45. J. Bruce Beckwith to Alice Dreger, June 19, 2000.
46. Alice Dreger to J. Bruce Beckwith, August 1–8, 2000.
47. J. Bruce Beckwith to Alice Dreger, June 19, 2000.
48. Alice Dreger to J. Bruce Beckwith, August 1–8, 2000.
49. Alice Dreger to Stella Mason, March 29, 2000.

5. The Future of Anatomy

1. Recent medical literature has contained a number of reports of prenatal detection of conjoinment, followed by elective abortion. See, for example, J. R. Wax et al., "Ultrasonographic Diagnosis of Thoracopagus Conjoined Twins in a Monoamniotic Triplet Gestation," *American Journal of Obstetrics and Gynecology,* 181, no. 3 (September 1999): 755–756; and F. Bonilla-Musoles et al., "Early Diagnosis of Conjoined Twins Using Two-Dimensional Color Doppler and Three-Dimensional Ultrasound," *Journal of the National Medical Association,* 90, no. 9 (September 1998): 552–556.
2. See April Herndon, "Carrying the Torch: Fatness and Nation in the Age of Weight Loss" (Ph.D. dissertation, Michigan State University, 2003).

3. Carl Elliott, *Better Than Well: American Medicine Meets the American Dream* (New York: W. W. Norton, 2003).

4. For an exploration of the possible conflicts between prenatal testing and disability rights, see Adrienne Asch, "Disability, Equality, and Prenatal Testing: Contradictory or Compatible?" *Florida State University Law Review*, 30, no. 2 (Winter 2003): 315–342. See also Erik Parens and Adrienne Asch, eds., *Prenatal Testing and Disability Rights* (Washington, D.C.: Georgetown University Press, 2000); and Allen Buchanan et al., *From Chance to Choice: Genetics and Justice* (Cambridge: Cambridge University Press, 2000). For a personal account of an incorrect prenatal diagnosis of club foot, and an intelligent analysis of the problems with ultrasounds, see Natalie Angier, "Ultrasound and Fury: One Mother's Ordeal," *New York Times*, November 26, 1996, p. C1.

5. Michel Foucault, *Birth of the Clinic*, trans. A. M. Sheridan Smith (New York: Pantheon, 1973).

6. Ronald Bayer, *Homosexuality and American Psychiatry: The Politics of Diagnosis* (Princeton: Princeton University Press, 1987). For a more extensive history of biomedicine's treatment of gay men and lesbians, see Jennifer Terry, *An American Obsession: Science, Medicine, and Homosexuality in Modern Society* (Chicago: University of Chicago Press, 1999).

7. For a radical reinterpretation of PMS, see Emily Martin, *The Woman in the Body: A Cultural Analysis of Reproduction* (Boston: Beacon Press, 1992), ch. 7.

8. Lori Schappell, personal communication, December 9, 2002.

9. Joseph P. Shapiro, *No Pity* (New York: Three Rivers Press, 1993), p. 332.

10. Kathi Wolfe, "Ordinary People: Why the Disabled Aren't So Different," *Humanist*, 56 (November–December 1996): 31.

11. Shapiro, *No Pity*, p. 112.

12. For a classic real-life example of the ways in which disability is socially constructed, see Nora Ellen Groce, *Everyone Here Spoke Sign Language: Hereditary Deafness on Martha's Vineyard* (Cambridge, Mass.: Harvard University Press, 1985).

13. Samuel Cartwright, "Diseases and Physical Peculiarities of the Negro Race" (1851), reprinted in John Harley Warner and Janet A. Tighe,

eds., *Major Problems in the History of American Medicine and Public Health* (Boston: Houghton Mifflin, 2001): 103.

14. N. Williams, "A Dissertation on 'Female Physicians'" (1850), reprinted in Warner and Tighe, *Major Problems,* p. 132. For a history of attempts to articulate the "natural" impairments of womanhood, see Cynthia Eagle Russett, *Sexual Science: The Victorian Construction of Womanhood* (Cambridge, Mass.: Harvard University Press, 1989); and Londa Schiebinger, *The Mind Has No Sex? Women in the Origins of Modern Science* (Cambridge, Mass.: Harvard University Press, 1989), chs. 6–8.

15. For a spirited exploration of the line dividing humans and other animals, see Jonathan Marks, *What It Means to Be 98 Percent Chimpanzee* (Berkeley: University of California Press, 2002).

16. For a discussion of the problems of using racial categories in medicine, see R. Witzig, "The Medicalization of Race: Scientific Legitimation of a Flawed Social Construct," *Annals of Internal Medicine,* 125, no. 8 (October 15, 1996): 675–679.

17. See Janice Hopkins Tanne, "Free at Last," *New York Magazine,* November 15, 1993, pp. 54–62.

18. Cartwright, "Disease and Physical Peculiarities," p. 105.

19. See, for example, Bruce E. Wilson and William G. Reiner, "Management of Intersex: A Shifting Paradigm," *Journal of Clinical Ethics,* 9, no. 4 (Winter 1998): 360–369; Philip A. Gruppuso, "Should Cosmetic Surgery Be Performed on the Genitals of Children Born with Ambiguous Genitals?" *Physician's Weekly,* August 16, 1999; Jorge Daaboul and Joel Frader, "Ethics and the Management of the Patient with Intersex: A Middle Way," *Journal of Pediatric Endocrinology and Metabolism,* 14, no. 9 (November–December 2001): 1575–1583; Justine Marut Schober, "A Surgeon's Response to the Intersex Controversy," *Journal of Clinical Ethics,* 9, no. 4 (Winter 1998): 393–397; Sarah Creighton and Catherine Minto, "Managing Intersex," *British Medical Journal (Clinical Research Edition),* 323, no. 7324 (December 1, 2001): 1264–1265.

20. See, for example, Catherine L. Minto et al., "The Effect of Clitoral Surgery on Sexual Outcome in Individuals Who Have Intersex Conditions with Ambiguous Genitalia: A Cross-Sectional Study," *Lancet,* 361 (April 12, 2003): 1252–1257.

21. Wendy Mouradian, "Deficits versus Strengths: Ethics and Implications for Clinical Practice and Research," *Cleft Palate Craniofacial Journal,* 38, no. 3 (May 2001): 255–559.

22. Rosemarie Garland Thomson, *Extraordinary Bodies: Figuring Physical Disability in American Culture and Literature* (New York: Columbia University Press, 1997), p. 137.

23. The Supreme Court's recent decisions on the Americans with Disabilities Act have weakened protections against disability discrimination. For a critical analysis, see Andrew J. Imparato, "The 'Miserly' Approach to Disability Rights," in Herman Schwartz, ed., *The Rehnquist Court: Judicial Activism on the Right* (New York: Hill and Wang, 2002).

24. For an excellent critique of the idea of "normal" in medicine, see Phillip V. Davis and John G. Bradley, "The Meaning of Normal," *Perspectives in Biology and Medicine,* 40, no. 1 (Autumn 1996): 68–76.

Acknowledgments

Many individuals and institutions have supported the projects that became this book. I am especially grateful to the staff, students, and faculty of the Lyman Briggs School at Michigan State University for providing me with an enclave of learning and goodwill. I benefited particularly from having as my students Alric Hawkins, April Herndon, Siavash Jabbari, Ruta Sharangpani, and Jennifer Zien. My research assistant Colleen Kiernan found and organized key materials, proofread on short notice, effectively argued with me over key points, and even recruited her grandmother to serve as a clipping service. She deserves much credit for the stronger aspects of this book. My colleagues at Lyman Briggs and the Center for Ethics and Humanities in the Life Sciences provided companionship and scholarly help; those I would particularly like to thank are Libby Bogdan-Lovis, Howard Brody, Kathie Ellis, Doug Luckie, Robert Shelton, Christie Tobey, and Tom Tomlinson. I am also indebted to Diane Ebert-May, Ed Ingraham, George Leroi, Steve Spees, and Elizabeth Simmons, administrators at my university who provided me with funding and leave time, and who understood how this project speaks to our land-grant mission. Michigan State University's Intramural Research Grants Program generously funded expenses related to the production of the book manuscript.

For the past several years I have been privileged to be a member of two long-term collaborative study groups central to this work, most recently the Hastings Center Surgically Shaping Children Working Group, a project led by Erik Parens and funded by the National Endowment for the Humanities.

Erik's intelligence, compassion, and belief in the importance of this work served as a model I have striven to follow. I am also very grateful to other core members of that group, including Priscilla Alderson, Adrienne Ash, Cassandra Aspinall, Dena Davis, James Edwards, Ellen Feder, Joel Frader, Art Frank, Lisa Abelow Hedley, Eva Kittay, Jeff Marsh, Paul Miller, Wendy Mouradian, and Hilde Lindemann Nelson. Earlier in my career I benefited enormously from conversations with Françoise Baylis, Tod Chambers, Carl Elliott, David Gems, Kathy Glass, Laurence Kirmayer, and Margaret Lock, the core members of the Enhancement Technologies and Human Identity Working Group, a project funded by the Social Sciences and Humanities Research Council of Canada.

Many other people also helped me to articulate the questions and answers central to this book. I am particularly indebted to Natalie Angier, Michael Barilan, J. Bruce Beckwith, Nancy and Mohammad Behforouz, Laura Beil, Danny Black, Margaret Carney, Cheryl Chase, Frank and Mary Dreger, Paul Dreger, Jeffrey Eugenides, Laura Ferguson, Carrie Fleig, Denise Grady, Phil Gruppuso, Debbie Hartman, Thea Hillman, Barron Lerner, Simi Linton, Angela Moreno Lippert, Dorothy Luckie, Margo Miles-Carney, Sarah Mitchell, James O'Neill, Bill Reiner, Lori Schappell, Nancy Segal, Ruta Sharangpani, Bea and Lynn Sousa, Volker Stollorz, Mark Stringer, Ellen Weissbrod, Bruce Wilson, Gretchen Worden, and members of the Intersex Society of North America, on whose board I have been honored to serve since 1998. I am also thankful to Adrienne Asch, Maria Ascher, Jeff Marsh, Rosemarie Garland Thomson, and two anonymous reviewers for Harvard University Press for reviewing various versions of this manuscript and providing substantial feedback. Assistance with illustrations was kindly provided by Jan Bondeson, Elaine Challacombe, Christopher Dreger, Laura Ferguson, Steve Wewerka, and John White.

Ann Downer-Hazell, my editor at Harvard University Press, amazed me throughout this process with her fortitude, kindness, good sense, and high standards (moral and editorial). I could not ask for more or better.

Finally, I am thankful to my husband, Aron Sousa, the *sine qua non* of my life.

Credits

1. Reproduced from a broadside in the Chang and Eng Bunker Papers, catalogue number OP-3761/1, Southern Historical Collection, Wilson Library, University of North Carolina at Chapel Hill.
2. Reprinted from George M. Gould, *Anomalies and Curiosities of Medicine* (Philadelphia: W. B. Saunders, 1897). Courtesy of the Wangensteen Historical Library, University of Minnesota.
3. Illustration by Christopher Dreger. Copyright © Christopher Dreger, 2003.
4. Reprinted from J. Bland Sutton and Samuel G. Shattock, "Report on a Living Specimen of Parasitic Fetus," *Transactions of the Pathological Society of London*, 39 (1887–1888): plate 37. Courtesy of the Wangensteen Historical Library, University of Minnesota.
5. Photograph by Steve Wewerka, Wewerka Photo. Reproduced by permission.
6. Reproduced from a broadside in the Chang and Eng Bunker Papers, catalogue number OP-3761/3, Southern Historical Collection, Wilson Library, University of North Carolina at Chapel Hill.
7. Reproduced with the permission of the Hospital for Sick Children, Toronto.
8. Copyright © AORN, Inc., 2170 South Parker Road, Suite 300, Denver, Colorado 80231.
9. Drawing by Mr. Devis, from the private collection of Jan Bondeson. Reproduced with the permission of Jan Bondeson.

Index

Abortion, elective, 86, 109, 117, 142, 143, 162n30, 184n1
Alderson, Priscilla, 74
Americans with Disabilities Act (ADA), 15, 129, 135, 187n23
Anatomy: social meanings of, 1–6, 8, 9–10, 12–16, 50, 61, 77, 117–118, 140–141, 142–155; legal treatments of, 3, 10, 14, 48–49, 84, 85–86, 95, 97–106, 108–109, 147, 148, 149–150; future of, 9–10, 111–112, 140–141, 142–155. *See also* Normal; Normalization; Sex role restrictions; Social justice; Unusual anatomies
Annas, George, 85, 93, 102, 104–105
Asch, Adrienne, 78, 129, 168n112
Aspinall, Cassandra, 153
Attard, Gracie, 84, 86, 95–108, 111, 178n51
Attard, Michaelangelo, 95–100, 105–107, 108
Attard, Rina, 95–100, 105–107, 108
Attard, Rosie, 84, 86, 95–106, 111, 178n51

Beckwith, J. Bruce, 137–140
Bianchi, Adrian, 84, 96, 106–107

Biddenden Maids, 46
Bijani, Ladan, 7, 41–43, 46, 66–67, 170n29, 171nn30,31, 174n69
Bijani, Laleh, 7, 41–43, 46, 66–67, 170n29, 171nn30,31, 174n69
Black, Danny, 2, 4, 134–135, 140
Blazek, Josepha, 40
Blazek, Rosa, 40
Blindness. *See* Vision impairment
Bogdan, Robert, 123
Breast-feeding, 17, 19
Bryant, Jessie Bunker, 25–26, 27
Bunker, Adelaide, 18, 21–22, 27, 62–63
Bunker, Chang, 17–19, 19–25, 27, 31, 37, 38–40, 43, 46, 49, 62–63, 68, 115, 120, 134
Bunker, Eng, 17–19, 19–25, 26, 27, 31, 37, 38–40, 43, 46, 49, 62–63, 68, 115, 120, 134
Bunker, Sallie, 18, 21–22, 25, 26, 27, 62–63
Byrne, Charles, 114–117, 118–120, 134, 136–141

Cady, Marlene, 33–35, 76
Cady, Ruthie, 33–35, 40, 46

Cady, Verena, 33–35, 40, 46
Carson, Benjamin, 66
Cartwright, Samuel, 148, 151
Chambers, Tod, 135–137, 140
Chase, Cheryl, 2, 4, 11–13, 14, 15, 128, 140
Circumcision, 173n66. *See also* Female genital mutilation
Civil rights movement, 16, 147, 148, 149, 150, 151, 154. *See also* Race
Cleft lip, 6, 16, 47, 58, 70–71, 147, 153–154, 160n3
Cleft palate, 56, 58, 70–71, 153–154, 160n3
Confidentiality (doctor-patient), 127–128
Conflicts of interest. *See* Medical professionals, conflicts of interest; Parents, identity crises
Conjoinment: frequency of, 6, 27, 29, 31, 73, 162nn27,30, experience of, 7, 16, 17–50, 67–68, 73, 74–75, 78, 85, 94–95, 111, 117, 129, 130, 132, 142, 155; causes of, 27, 29, 161n26; types of, 28–30. *See also* Individuality; Marriage, of conjoined twins; Parasitic conjoinment; Sacrifice surgeries; Separation surgeries; Sexuality of conjoined twins; "Siamese twins"; Singleton assumption
Consent to surgery. *See* Sacrifice surgeries, consent to; Separation surgeries, consent to
Cosmetic surgery. *See* Normalization

Dickson, Alan, 84, 106, 107
Disability: meanings of, 10, 78, 98, 117, 124, 126, 128–130, 133, 134–135, 139–140, 146–148, 154–155, 182n23, 185n12, 187n23; rights movement, 10, 16, 73, 78, 129, 146–148, 154–155, 163n33, 164n52, 185n4
Doctors. *See* Medical professionals
Double-X syndrome, 14
Dreger, Paul, 14–15
Du Bois, W. E. B., 9
Dwarfism: social aspects, 1, 4, 5, 6, 58, 113, 149; experience of, 4, 16, 56, 117, 129, 134–135, 173n57

Eugenics, 113, 117–118, 143
Evidence-based medicine. *See* Normalization, effectiveness of surgeries; Separation surgeries, outcome data (and lack thereof); Uncertainty in medicine

Fat (as source of stigma), 21, 47, 72, 117, 143
Female genital mutilation (also known as female circumcision), 13–14
Ferguson, Laura, 130, 131, 140
Fishman, Steven, 110
Frank, Arthur, 56
Freak shows, 8, 27, 123, 125–127, 138, 181n7. *See also* Unusual anatomies, displays and representations of

Gay and lesbian rights movement, 145, 147, 149, 150, 151, 185n6
Gender. *See* Anatomy, social meanings of; Double-X syndrome; Gay and lesbian rights movement; Intersex; Sex role restrictions; Women's rights movement
Genital normalization surgeries, 5, 11–14, 53, 58, 61–62, 64, 68, 77, 101, 107–108, 149, 152–153, 172n50, 173n66. *See also* Circumcision; Female genital mutilation; Sex reassignments on male conjoined twins

Giantism, 6. *See also* Byrne, Charles

Gibb, Margaret, 40, 46

Gibb, Mary, 40, 46

Goh, Keith, 41–42, 171n30

Green, Joanne, 47

Hartman, Debbie, 72–73

Hensel, Abigail, 36–37, 38, 40, 41, 43, 44, 61–62, 129

Hensel, Brittany, 36–37, 38, 40, 41, 43, 44, 61–62, 129

Hensel, Mike, 37, 61, 76

Hensel, Patty, 5–6, 37, 61, 76, 129

Hermaphroditism. *See* Intersex

Heroism and bravery (attribution to people with unusual anatomies and other disabilities), 43, 59, 108, 126, 130, 147, 155, 182n23

Hilton, Daisy, 9, 48–49

Hilton, Violet, 9, 48–49, 63

Holton, Eilish, 51–55, 58, 60, 65, 125, 128

Holton, Katie, 51–55, 60, 65, 125–126, 128

Holton, Liam, 52, 53, 60, 126

Holton, Mary, 52, 53, 60

Htut, Lin, 37–38, 69

Htut, Win, 37–38, 69

Individuality, 6–7, 18–19, 31–33, 40–41, 42, 50, 63, 88, 132, 143, 155, 163n33. *See also* Conjoinment, experience of; Singleton assumption

Individuation, 22, 32, 35, 37, 40–41, 44–46, 67, 74–75, 165n73

Informed consent. *See* Sacrifice surgeries, consent to; Separation surgeries, consent to

Intersex, 6, 10–16, 56, 57, 58, 71, 143, 147, 149, 150; experience of, 2, 4, 11, 13, 62, 68, 72–73, 138, 147, 152,

173n63, 174n72; rights movement, 13, 147, 151–153; images of, 118. *See also* Genital normalization surgeries

Jones, Willa, 35

Lakeberg, Amy, 84, 86–95, 111

Lakeberg, Angela, 84, 86–95, 111

Lakeberg, Kenneth, 86–90

Lakeberg, Reitha ("Joey"), 86–89

Laloo, 30

Lazareff, Jorge, 62

Lesbian and gay rights movement. *See* Gay and lesbian rights movement

"Manchester twins." *See* Attard, Gracie; Attard, Rosie

Manheim, Camryn, 47

Marriage: of conjoined twins, 18, 21, 24, 27, 31, 48–50, 61–63, 147, 171n31; as a metaphor for conjoinment, 132. *See also* Sexuality of conjoined twins

Marsh, Jeffrey L., 160n3

McCarther, Yvette, 35–36, 37

McCarther, Yvonne, 35–36, 37

McCoy, Christina, 36, 46, 63, 120–122, 165n73

McCoy, Millie, 36, 46, 63, 120–122, 165n73

Medicalization, 6, 8, 10, 24, 56, 58–59, 71, 75–76, 77, 80, 124–127, 128–130, 140, 142–146, 147, 148–149, 150, 154–155, 173n62

Medical professionals: personal experience with separation surgeries, 38, 42, 51–52, 84, 89, 93, 106, 110; understanding of own role, 57, 59, 62, 72, 75, 76, 77, 78, 84–85, 91–92, 100, 116–117, 120–124, 128–130, 136, 143, 144, 151, 183n28; conflicts of

Medical professionals *(continued)*
   interest, 75–76, 117, 124–125, 127,
   129, 137–141, 144, 174n69
Miles-Carney, Margo, 26–27
Mouradian, Wendy, 153–154
Muraskas, Jonathan, 91, 94
Museum specimens, 25, 27, 114–117,
   123–124, 134, 135–141, 161n25

Normal (variations in concept of), 7,
   9–10, 13, 14–16, 24, 31, 33, 37–38,
   42, 43–44, 46, 47, 52, 55, 56, 57, 58,
   60–61, 63, 67, 68, 72–73, 75–76, 78,
   81–82, 93, 95, 97, 99, 101, 104, 107,
   117–118, 129–130, 142–155,
   169n20, 187n24
Normalization: motivations for surgeries, 3–4, 5, 6, 7–8, 10, 11, 12, 55–
   57, 58–59, 74–82, 124–125, 142–143,
   144–145, 149, 173n62, 174n72; nonsurgical, 3–4, 7–8, 10, 17, 55, 58,
   132, 142–145; effectiveness of surgeries, 7–8, 11–12, 16, 56, 60–66,
   70–73, 74, 77, 80–81, 153; ethics of,
   7–8, 10–16, 51–82, 144, 150–151,
   152–154, 173n66. *See also* Genital
   normalization surgeries; Sacrifice
   surgeries; Separation surgeries
Normate, definition of, 158n13
Nurses. *See* Medical professionals

O'Neill, James, Jr., 88, 94

Parasitic conjoinment, 30, 85, 94, 101–
   102, 108–109, 114, 176n13
Parents: experience with children with
   unusual anatomies, 5–6, 21, 33–35,
   37, 42–43, 47, 55–57, 71, 72–73, 76–
   78, 80–81, 84, 86–88, 92, 93, 96–98,
   105, 107, 108, 109–111, 144, 146,

   153, 154–155, 167n111, 170n23; be-
   haviors towards children with
   unusual anatomies, 19, 33–35, 36–
   37, 45, 52–53, 55–57, 71, 75–78, 86–
   88, 89–90, 96, 98, 105–108, 109, 111,
   117, 126, 160n3; identity crises, 57,
   59, 75, 77, 98, 102, 170n23
Peer support. *See* Support by peers
Performance. *See* Unusual anatomies,
   displays and representations of
Pity, 5–6, 19, 34, 47, 130, 135, 147
Psychology of conjoinment. *See*
   Conjoinment, experience of; Individuality; Individuation;
   Psychosocial support; Separation
   surgeries, outcome data (and lack
   thereof); Singleton assumption;
   Support by peers
Psychosocial support, 70–73, 74, 77–
   78, 80–81, 128–129, 130, 153. *See
   also* Support by peers

Quiej-Alvarez, Maria de Jesus, 62,
   170n28
Quiej-Alvarez, Maria Teresa, 62,
   170n28

Race (social meanings of), 9–10, 15–
   16, 21, 26–27, 47, 113, 115, 124,
   125–126, 148, 149, 181n2. *See also*
   Civil rights movement

Sacrifice surgeries, 8, 83–112; cases of,
   83–84, 86–111; definitions of, 83,
   84, 97; ethics of, 83–112; motivations for, 83, 84–85, 87–88, 91–95,
   97, 99–105, 108–109, 110, 111–112;
   frequency of, 84; arguments against,
   87, 90–95, 97–98, 100–101, 105, 109,
   111–112; consent to, 88, 92, 94–95,

105, 109, 110; outcomes of, 88–90, 106–108, 111–112

Schappell, Lori, 2, 4, 5, 40–41, 43, 48, 49, 130, 132–134, 147

Schappell, Reba, 2, 4, 40–41, 43, 130, 132–134

Separation surgeries: motivations for, 5, 7–8, 24, 41–43, 53, 55–57, 60–63, 64, 65, 71–72, 73–82, 120, 160n13, 170n29; cases of, 7, 38, 41–43, 51–55, 62, 64–65, 66–70, 75, 78–79; consent to, 7, 34–35, 41, 60, 66–67, 73–75, 76–78, 80–81, 126, 151; decisions against, 7, 24, 33, 34, 35, 36–37, 43–44, 45–46, 67–68, 73–75, 75–78, 80–82, 116n92; ethics of, 7–8, 42, 51–82; outcome data (and lack thereof), 62, 63–71, 73–74, 80–81. *See also* Sacrifice surgeries

Sex reassignments on male conjoined twins, 68–70

Sex role restrictions, 1, 2, 3, 12–14, 148–149. *See also* Intersex; Sexuality of conjoined twins; Women's rights movement

Sexuality of conjoined twins, 24, 31, 49–50, 61–63, 68–70, 99, 101, 107–108, 121–123. *See also* Marriage, of conjoined twins

Shame: as response to unusual anatomies, 5, 6, 11, 55, 71–72, 78, 133–134, 144, 147; as response to normalization surgeries, 11, 71–72; as motivation for normalization, 58, 71–72, 144; and coming out, 126–127, 154–155, 184n42. *See also* Stigma

Sharangpani, Ruta, 2, 4, 5, 47

Shrestha, Ganga, 41, 70

Shrestha, Jamuna, 41, 70

"Siamese Twins," 22, 115, 159n1

Side Show, 49–50

Singleton, my use of the term, 6–7

Singleton assumption, 6–7, 18, 31, 60, 63, 64, 67, 68, 78–80, 103–104, 108–109, 125, 155

Smith, J. David, 40

Social justice, 2–3, 6, 8, 14–16, 72, 78, 87, 90–92, 116–117, 120, 123–124, 129, 145–155. *See also* Civil rights movement; Gay and lesbian rights movement; Intersex, rights movement; Women's rights movement

Soto, Darielis Milagro, 86, 109–112

Soto, Ramon, 109–111

Soto, Sandra (mother), 109–111

Soto, Sandra Ivellise (daughter), 86, 109–111

Spitz, Lewis, 52, 65, 68–69, 107, 128

Staring and mockery, 6, 55, 61, 133. *See also* Pity; Stigma

Stigma, 2–5, 11–12, 14–15, 16, 21, 26–27, 55, 60–61, 64, 70, 71–72, 78, 82, 127–130, 132, 140, 143, 144–146, 157n5. *See also* Pity; Shame; Staring and mockery

Stringer, Mark, 51–52, 54, 55, 65–66

Support by peers, 72–73, 77, 137, 153, 154–155, 173nn57,63

Surgeons. *See* Medical professionals

Surgeries. *See* Genital normalization surgeries; Normalization; Sacrifice surgeries; Separation surgeries

Talk shows, 126–127

Taveras, Carmen, 7

Taveras, Rosa, 7

Thomson, Rosemarie Garland, 154, 158n13

Two-headed boy of Bengal, 114

Uncertainty in medicine, 57, 59, 63–
    64, 65–66, 70, 80, 81, 111–112, 153
Unusual anatomies: standard stories
    about, 2, 6, 8, 12, 16, 47, 62–63,
Unusual anatomies *(continued)*
    75–76, 111–112, 123, 124–126, 138,
    147, 148–149, 174n72; alternative
    stories about, 6, 8, 14, 15–16, 47,
    75–76, 126–127, 128–135, 138–141,
    147–148, 149, 150–155, 159n26; dis-
    plays and representations of, 8, 15–
    16, 18, 21, 22–24, 25, 27, 35, 36, 45,
    48, 70, 75, 80, 87–88, 98, 100, 108,

113–141, 174n69, 183n27; experi-
    ence of congenital vs. acquired,
    47, 167n111. *See also* Singleton
    assumption

Vision impairment, 2, 4, 16, 47, 148

Weight and weight loss. *See* Fat
Weissbrod, Ellen, 50, 130, 132–133,
    137
Women's rights movement, 14, 147,
    148–149, 150, 151, 154, 155